DESIGN OF
HIGH TEMPERATURE METALLIC COMPONENTS

Proceedings of a seminar held at the

JOINT RESEARCH CENTRE
of the
COMMISSION OF THE EUROPEAN COMMUNITIES
Petten Establishment, The Netherlands

and organised by the

CEC High Temperature Materials
Information Centre,
Petten, The Netherlands

DESIGN OF HIGH TEMPERATURE METALLIC COMPONENTS

Edited by

R. C. HURST

Joint Research Centre, Petten Establishment, The Netherlands

ELSEVIER APPLIED SCIENCE PUBLISHERS
LONDON and NEW YORK

ELSEVIER APPLIED SCIENCE PUBLISHERS LTD
Ripple Road, Barking, Essex, England

Sole Distributor in the USA and Canada
ELSEVIER SCIENCE PUBLISHING CO., INC.
52 Vanderbilt Avenue, New York, NY 10017, USA

British Library Cataloguing in Publication Data

Design of high temperature metallic components.
1. Metals at high temperatures
I. Hurst, R. C. II. CEC High Temperature Materials Information Centre
620.1′1217 TA460

ISBN 0-85334-301-2

WITH 99 ILLUSTRATIONS

Publication arrangements by: Commission of the European Communities, Directorate-General for the Information Market and Innovation, Luxembourg

EUR 9159 EN

Printed in Great Britain by Galliard (Printers) Ltd, Great Yarmouth

Preface

The successful operation of metallic alloys under conditions prevalent in high temperature plant such as gas turbines, heat exchangers, coal gasifiers, and chemical or petrochemical process units is completely determined by the information available to the designer in the form of material specifications, material properties and the available codes he has to work with. However, the plant operating conditions are so varied, and also often extremely aggressive including very high temperatures, corrosive elements, erosive particles, thermal cycling and high loads, that these are not systematically or even easily incorporated into design codes.

The purpose of the discussion seminar 'Design of High Temperature Metallic Components' organised by the Information Centre of the High Temperature Materials Programme of the European Commission Joint Research Centre, Petten Establishment, was both to guide scientists towards the information the high temperature designer requires and conversely to inform the design engineer of the present state of materials research in relation to high temperature component design.

The lectures were chosen so as firstly to review available knowledge concerning high temperature design codes and practices appertaining to two important industries: electrical generation and chemical/petrochemical production. Although the paper on the latter subject had somewhat less a review characteristic, many of the important problem areas were highlighted. Secondly, the methods and progress in metallurgical research aimed at producing more directly usable data and the possibilities for component lifetime prediction were thoroughly delineated in two successive lectures. These were followed by a detailed description of the engineer's contribution to the component design problem via exact finite element analysis methods. Finally, the complex situation was covered where property interactions, complex stresses and component testing require that a combined interdisciplinary materials/engineering knowledge be applied.

It is hoped that the papers presented during the discussion seminar, in book form, should satisfy both engineers and materials scientists making their individual contributions to improving high temperature metallic component design and, perhaps more importantly, bring them closer together so that they can synergistically reach their goal.

As scientific co-ordinator (R.C.H.) and Information Centre Head (M.M.) it is a pleasure for us to thank C. H. A. Townley, Ph. Holl, R. W. Evans, B. Wilshire, D. R. J. Owen, O. J. A. Gonçalves, and F. Schubert and his co-authors for their efforts in both presenting and preparing an evidently high-quality collection of papers and to the delegates at the seminar who contributed to its success through their active participation.

R. C. HURST
M. MERZ

Contents

Introduction

M. VAN DE VOORDE

Programme Manager, High Temperature Materials Programme, Joint Research Centre, Petten Establishment, The Netherlands

The design of engineering components for service at elevated temperatures is a great deal more complicated than that for equivalent components operating at ambient temperature, because various degradation processes become active and are much accelerated by rise of temperature. The processes contributing to the progressive fall in serviceability of a component are of three main types: first, structural changes in the constitution of the alloys; second, deformation and crack growth under the influence of the imposed mechanical stresses; and third, chemical corrosion of the surface of the component by reaction with the environment. Further, to complicate the situation, these three processes interact with one another, so that while they may be studied separately under controlled laboratory conditions, the effects on components in service are less readily separated.

The design of a specific high temperature component has, therefore, to be made in the light of a body of property data which has, for the most part, been determined on idealised specimens tested under simplified conditions of temperature profile, stress cycle and environment. The component is then expected to perform satisfactorily in service when all the controlling factors referred to may differ significantly from those of the available data. The time of operation may also be different from that of the test data. Thus, to allow for the uncertainties involved, generous safety factors are normally applied, but while these may ensure safe operation, they can involve wasteful use of materials. There is therefore a clear incentive to combine safety in design with economical use of materials and, in essence, this forms the main theme of this seminar.

In addition to the interaction between the effects of imposed stress and corrosion, which can be studied reasonably satisfactorily on simple samples, there are other important factors which arise when considering the application of standard data to component design. These include the effects of size and geometry, and the problems of joining.

Most data are derived from samples with a ruling section of the order of 1 cm, while components may be much larger and may be produced by a significantly different process leading to differences in both macrostructures and microstructures.

Larger sizes also lead to increased internal stresses resulting from heat treatments, which, even if relieved by thermal relaxation in the early stages of service, cause redistribution of stress patterns. Thermal stresses generated during change of service condition are also dependent on the component geometry, and, of course, have to be allowed for in operating procedures.

Some high temperature components, such as turbine blades, are integral parts in which the critical areas are remote from attachment points, and are small enough to allow simulated service testing to confirm design. However, others, such as steam pipes or reformer tubes, require joints and fixtures within the critical zones, and here the effects of welding or mechanical joining have to be considered, including the influence of *in-situ* stress-relieving treatments.

While interpolation of the effects of temperature or of stress on time-dependent failure mechanisms such as creep and fatigue is usually reliable, extrapolation of data beyond the range of observation is notoriously risky. Many attempts have been made to derive fundamental or phenomenological formulae to aid the process, but fully reliable success has not been achieved. Not only does this imply that fully reliable designs for a 30 years life require full-time test data but also that, even if these data are available, the material is 30 years old. Almost certainly it is not representative of that from modern production processes.

The influence of these and other factors demanding close attention to the insurance of safety in operation, combined with economy in construction, points to the need for the testing of more advanced types of specimen or model component under conditions simulating as closely as possible those anticipated in service. With their increasing complexity and size such facilities are expensive and it seems necessary that co-operative action to provide them is required.

At this seminar, we have the benefit of presentations from the representatives of major industries involved in large-scale high temperature plant and of those concerned with basic studies of materials property data and their application to design. It is hoped that the lectures presented and the ensuing discussion will prove helpful and stimulating to those engaged in this field.

1

Design Methods for High Temperature Power Plant Structures

C. H. A. Townley
Central Electricity Generating Board, Berkeley Nuclear Laboratories, Berkeley, UK

1. INTRODUCTION

As an introduction to the seminar, I have been asked to provide a review of the design methods and design criteria currently in use for both nuclear and fossil fuelled power plant.

It would be an impossible task to compare and contrast the very many national design codes which are available, especially for conventional power plant and pressure vessels. Instead, I have chosen to examine two examples—BS 1113,[1] which is representative of the design codes employed for power station boiler plant, and ASME Code Case N47,[2] which is being developed for high temperature nuclear reactors, especially the liquid metal fast breeder reactor. The first derives from an industry with many years of operating experience behind it; the second ventures into a relatively unknown field, where procedures must necessarily be derived from first principles.

In accordance with the theme of the seminar, I have been asked to draw attention to the topics which, as an engineer, I believe should be of interest to metallurgists. However, the materials issues cannot be divorced from the design procedures which are adopted. The designer of a pressure vessel or a nuclear component is faced with a structure of complex geometries, containing stress concentrations and stress gradients. Often it will be subject to multiaxial stresses, and the loading may be far from constant. This is in contrast to the simple uniaxial tests which are employed to derive the materials data which the designer must use. The art of successful design is to ensure that stress concentration features, the welds used in fabricating the component, the loadings applied in service and so on do not lead to premature failure.

TABLE 1
TYPICAL UK POWER PLANT MATERIALS

Power plant	*Component*	*Material*	*Approximate operating temperature (°C)*
500 MW power station boiler	Superheater tubes	Mild steel, 1Cr1Mo, $2\frac{1}{4}$Cr1	370 min
		Eshette, AISI 347, 316, 304	620 max
	Reheater tubes	Mild steel, 1Cr1Mo, $2\frac{1}{4}$Cr1 Eshette, AISI 347, 316, 304	400–620 max
	Superheater headers	AISI 316 or 347	370–580
Advanced gas cooled reactors	Evaporator/ superheater tubes	C Steel, 9Cr1Mo Austenitic AISI 316	250–565
	Reheater tubes	AISI 316	400–565
	Superheater headers	AISI 316	150–565
	Primary circuit	AISI 304, 347, 316	150–565
Fast reactor (proposed design)	Primary circuit	AISI 316	400–575
	Secondary circuit	9Cr1Mo	350–525

I have therefore provided, for both examples, a brief introduction to the design principles which they employ. I cannot, in the short space available, discuss the finer points of the design procedures. To do so would only be justified in a conference of engineering specialists. However, I believe that what I have to say should be sufficient to illustrate the way in which the materials data are used, and will promote discussion on what further information on materials behaviour the designer needs.

One important point must be kept in mind. No design code provides a fully comprehensive procedure which, followed through step by step by a newcomer, would enable a satisfactory pressure vessel to be designed and manufactured. That is not what is intended. The codes essentially provide base-line methods and criteria, which are to be used by specialist organisations with relevant experience in the field.

Many factors which contribute to a successful design, which will give trouble-free service throughout its specified life, are not discussed at all in the codes, or at least receive scant attention. Some, such as the performance of welds at high temperature, will be considered in this paper. Others, such as corrosion, erosion, and environmental cracking are judged to be outside the scope of the present seminar, but nonetheless are of crucial importance.

Table 1 provides a summary of the materials most commonly used in high temperature power plant, and the temperatures at which they operate. I have drawn on UK experience for this table, and I should mention that alternative materials are used for similar duties in other countries.

2. DESIGN CODES FOR POWER STATION BOILERS

Design rules for chemical vessels and steam generating plant are based, to a large extent, on previous practice. There is a considerable background of information on what has given satisfactory performance in the past, and what is to be avoided in the future. At the same time, the codes are being continually updated in a way which takes advantage of improved knowledge in the fields of structural analysis, materials properties and failure criteria.

In the UK BS 5500[3] is the master code for unfired pressure vessels, BS 806[4] is used for power station pipework and BS 1113[1] for large steam boilers. There is a strong similarity in all three codes with regard to their basic principles, resulting from an integrated approach to the preparation of pressure vessel design rules in the British Standards Institution. I have chosen BS 1113 for discussion in the present paper.

This code provides a design route which avoids extensive stress analysis, although there is nothing to prevent the designer doing this if he so wishes. The basic membrane thickness of the vessel is fixed by a simple mandatory formula. Charts and formulae are provided to estimate the additional thickness needed in stress concentration regions. Design procedures are the same for components operating at relatively low temperatures and those operating where creep is important.

The minimum thickness for a component, such as a header, in the regions away from geometrical discontinuities is obtained from the formula

$$t = \frac{pD_i}{2f - p} \qquad \text{or} \qquad t = \frac{pD_o}{2f + p} \tag{1}$$

where $t =$ is the minimum thickness of shell, p is calculation pressure, D_i is the inside diameter of the shell, $D_o =$ is the outside diameter of the shell, and f is the design stress of the metal at the appropriate temperature. A joint efficiency factor may be required if the component contains welds.

Similar formulae are used to obtain the minimum thickness of straight tubes and integral pipework. Here the code requires an additional provision to be made for surface corrosion which may occur during the life of the plant. Allowance is also made for thinning which may occur at pipe and tube bends.

The design stresses to be used in these formulae are tabulated in the code. At high temperature, the design stress is the minimum stress to cause rupture in the required design life, divided by 1·3. In general, the rupture stresses are those agreed by the International Standards Organisation for the classes of steel permitted by the code. In some instances, where long-term UK experience shows that the ISO values are conservative, higher design stresses are permitted.

Materials are restricted to those which have given satisfactory performance in service. Chemical and physical properties are tightly specified.

As stated earlier, detailed stress analysis is not required for the various stress-raising features, such as branch connections, end closures, vessel supports and ligament regions. Instead, the code provides a set of simple rules to estimate the amount of thickening required in these regions to reduce the stress concentration to an acceptable level.

A full discussion on how each of the stress concentration regions is dealt with is outside the scope of the seminar. To illustrate the procedure, I have chosen as a typical example the information provided to the designer about the reinforcement needed in branch connections. A series of charts is provided, an example relating to protruding nozzles being shown in Fig. 1. T is the thickness of the main shell away from the discontinuity. T_r is the local thickness of the main shell to provide the reinforcement, and t_r is the local thickness of the branch pipe required to provide the reinforcement. ρ is the geometrical factor $d/D.\sqrt{(D/2T_r)}$. C is a factor, tabulated in the code, which takes account of any external loads which are applied to the connection.

Within certain specified limits the designer is permitted to put the reinforcement either in the main shell or in the branch pipe, or in a combination of both. It is a simple matter to read off from the chart what local thickening is required in the branch pipe and the main shell.

At first sight, it is perhaps surprising that the identical reinforcement rules can be applied to branch connections in high temperature plant as well as at lower temperatures. The explanation is provided from considerations of the way in which the design charts were derived.[5]

The starting point was a large amount of data, obtained both

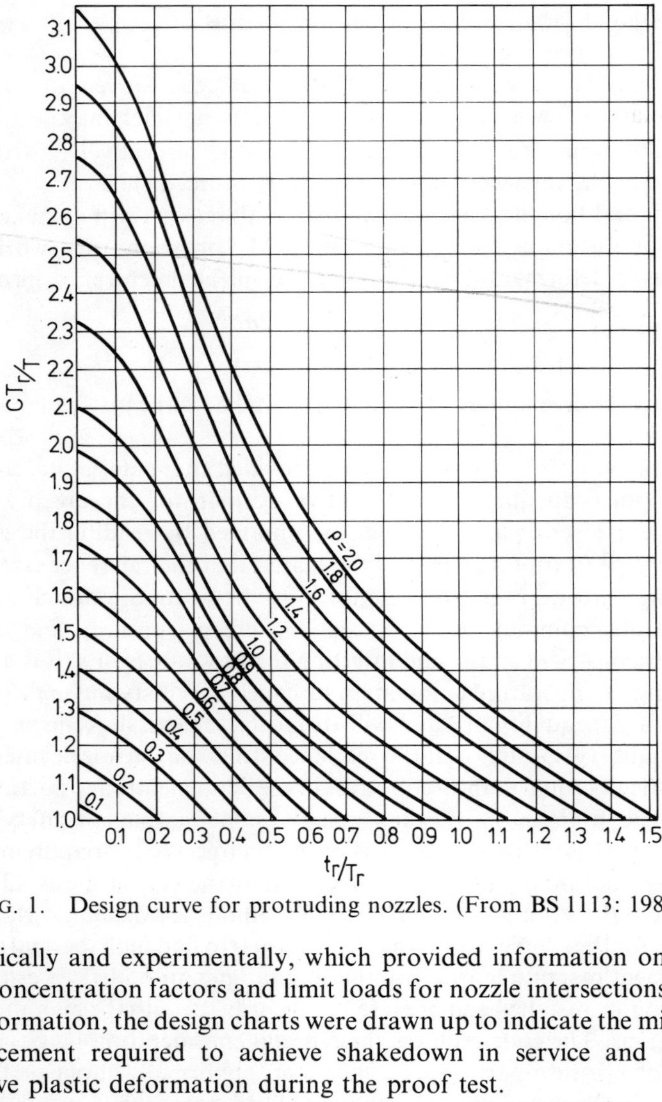

FIG. 1. Design curve for protruding nozzles. (From BS 1113: 1983[1].)

theoretically and experimentally, which provided information on elastic stress concentration factors and limit loads for nozzle intersections. From this information, the design charts were drawn up to indicate the minimum reinforcement required to achieve shakedown in service and prevent excessive plastic deformation during the proof test.

The first requirement, transposed into the more familiar ASME notation (see Section 3.1), is

$$P_L + P_B + Q \leq 2{\cdot}25S_m \tag{2}$$

The value of 2·25 was used when the UK rules were drawn up, in place of the more usual value of 3·0, to provide additional margins of safety.

The second requirement can be expressed as

$$\frac{(P_{\text{lim}})_{\text{intersection}}}{(P_{\text{lim}})_{\text{plain shell}}} \simeq 1 \tag{3}$$

where the numerator and denominator refer, respectively, to the limit pressure of the intersection and plain unreinforced shell.

Leckie and Ponter[6,7] have shown that, in the creep regime, shakedown is still an important concept. Using the ASME notation, their work shows that plastic deformation can be neglected in the creep range, provided

$$P_{\text{L}} + P_{\text{B}} + Q \leq \frac{2n}{n+1} S_{\text{y}} \tag{4}$$

where n is the creep index of the material and S_{y} is the instantaneous yield stress.

Goodall *et al.*[8,9] have shown that, provided the material has adequate creep-rupture ductility, a conservative estimate of the creep life of a component of complex shape can be obtained by reading the stress to rupture/time to rupture curve of the material at the reference stress level.

Taking into account the relationships between S_{m}, which for high temperature components is based on creep rupture, and S_{y}, the instantaneous yield stress, and inserting realistic values of n, it is apparent that a high temperature component designed to satisfy inequality (2) will also satisfy inequality (4), and will therefore achieve shakedown.

Inequality (3) ensures that the reference stress for the nozzle intersection is numerically equal to the hoop stress in the plain membrane portion of the vessel. The intersection will thus have the same margins of safety against creep rupture as the main body of the vessel, subject to the requirement that the creep-rupture ductility of the material at the end of life is adequate.

BS 1113 does not give specific design methods for dealing with thermal stresses, for the simple reason that it is unusual to find high thermal stresses, of sufficient magnitude to cause thermal fatigue, in boilers of established construction, installed and operated in accordance with the manufacturer's instructions. The code requires that special consideration be given to the design of pressure parts when, *inter alia*, abnormally rapid or frequent changes of pressure or temperature are likely to occur.

BS 1113 has little to say on the subject of strength of welds at high temperature. It is assumed that, provided the creep-rupture strength of the weld metal is similar to that of the parent material, the 'joint efficiency factor' can be taken as unity. In general, this assumption is borne out by the good performance which has been obtained in service with vessels designed to BS 1113. However, it must be remembered that this would probably not have been achieved in the absence of development work carried out by fabricators on high temperature performance of weld metals, research carried out by plant owners such as the CEGB, and collaborative investigations which have been undertaken, for example, by the ERA. The more recent research, such as is discussed in Ref. 10, and illustrated in Fig. 2, casts doubt on the adoption of matching strength as the most important criterion. It appears that ductility of all the parts of the complete weldment, including the heat affected zone (HAZ), is an important factor as is the relationship between the creep deformations of the various parts.

Welded transition joints occur in most modern power station boilers, between austenitic and ferritic tubes and pipes. No specific design rules are provided in BS 1113 for such joints between dissimilar metals. Again, satisfactory performance has been assured through research and development programmes carried out jointly between the manufacturers and large customers, such as the CEGB.

Experience with boilers designed to BS 1113 suggests that there are few, if any, deficiencies in the design procedures, or in the choice of material design stresses. In general, vessels designed for a nominal life of 100 000 h have achieved that life, and, as will be discussed below, attention is now focussed on demonstrating that these are capable of continuing in operation well beyond that time.

Where difficulties have been experienced with boiler plant in service, and these have been relatively few, the causes have usually been found elsewhere: temperatures actually higher than those allowed for in the design; extraneous system stresses in interconnecting pipework; and occasional deficiencies in quality control during manufacture.

It would thus appear that there is no pressing need for major alterations to BS 1113 and associated codes. However, it would be wrong to imply that no further improvements are required. For example, more precise guidance is desirable on thermal fatigue and, more generally, on the fatigue of attachment welds; design criteria for very thick cylindrical shells at high temperature could also be improved. Additional design charts for stress concentrating features will be provided as more results become available from theoretical and experimental stress analyses.

As far as high temperature data on the parent materials are concerned, some long-term tests remain to be completed to obtain creep rupture properties for times upwards of 100 000 h. The general impression is that present design stresses are over-conservative. It is pertinent to ask whether design values could be increased, without compromising safety, in view of

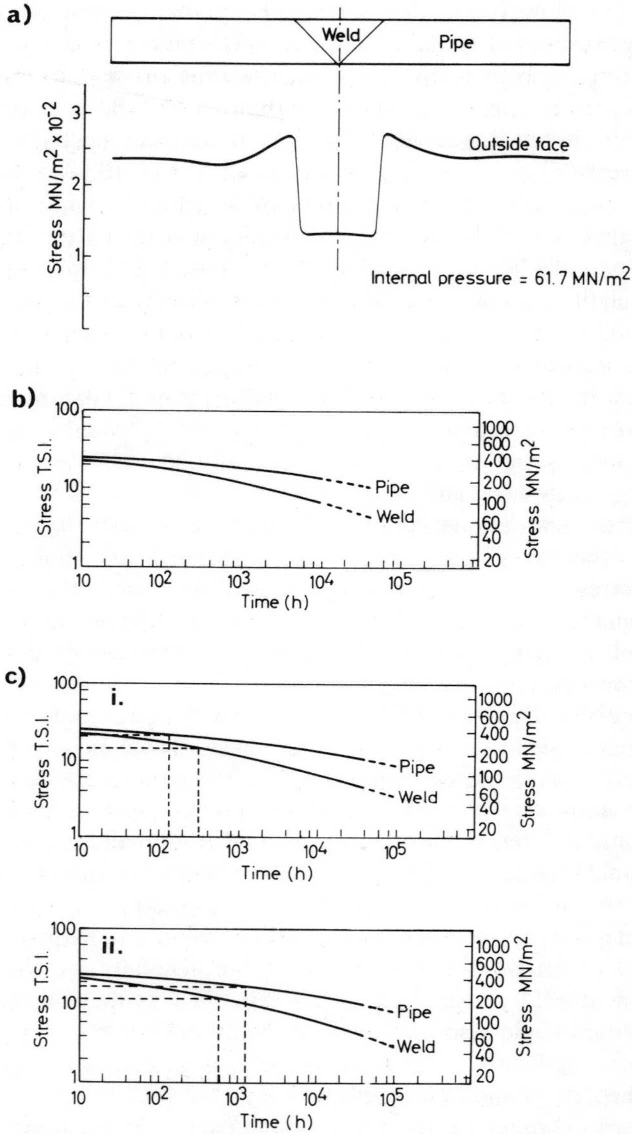

FIG. 2. Redistribution of stress in weld due to creep, and its effect on rupture life. (a) Off-loading of the tangential stress in the weld. (b) The stress–rupture properties of weld and parent material. (c) Effect of diverging stress–rupture curves on location of failure: (i) failure in pipe; (ii) failure in weld.

the very large remaining lives which have been estimated for plant which has already been in service for over 100 000 h. One possible explanation, which is currently being investigated within the CEGB, is that this conservatism arises from the nature of the conventional creep rupture tests. Oxidation reduces the metal section as the test proceeds, leading to premature failure. With the thick sections which are found in boiler plant any surface oxidation would have a negligible effect on stress levels, and thus the life would be considerably higher than predicted from the uniaxial data.[11]

Several of the earliest 500 MW units in the CEGB are now approaching 100 000-h operation, and are showing little sign of reaching the end of their life. There are clearly considerable economic benefits if such plant can be kept in operation into the foreseeable future, without replacement of major boiler components such as headers and drums. The tubing itself presents less of a problem, as this can be replaced relatively cheaply and without causing extended outages.

A considerable amount of effort is now being devoted within the CEGB to demonstrate continuing fitness for purpose of high temperature boiler components and to estimate their remaining life. Indeed, there is a growing general interest in the topic of life extension. Unfortunately, space does not permit a full description of the procedures at this seminar, and I would refer those who are interested to the recent British Standards' Published Document, describing the current state of the art in the UK.[12]

From a metallurgical point of view, the most interesting research is concerned with examining the change in the microstructure in the material to predict the amount of creep damage accumulated so far, and with tests on cut-out material to quantify the remaining life. There is also a body of work which is investigating crack growth phenomena at high temperature, so that appropriate decisions can be made about any defects which are revealed as a result of non-destructive examination of the components.

3. CODE CASE N47

I have chosen ASME Code Case N47[2] to illustrate the very much more complex analysis which is required for the design of nuclear plant which operates at high temperatures.

The primary circuit components of the earlier commercial reactors operated at relatively low temperatures. In the British Magnox plant, the maximum metal temperatures were a little under 400 °C. Pressure parts were designed to BS 1500,[13] supplemented by additional requirements agreed between the CEGB and the Independent Inspection Authorities. The advanced gas cooled reactors, which followed, operate at higher temperatures. The primary pressure vessel is of a prestressed concrete design, with a mild steel liner. The inner surface of the liner is thermally insulated from the hot carbon dioxide coolant, as are many of the internal structures. Special design rules apply to the prestressed concrete pressure vessel. The basis for the design of the internal structural components was BS 1500, BS 3915[14] and, later, BS 5500.[3] In all cases the code requirements were supplemented by agreement between the CEGB, the plant designers and the Independent Inspection Authorities. The steam generators, which are enclosed within the concrete pressure vessel, were generally to BS 1113, again adapted as necessary. Code Case N47 has been used to assess a few of the internal components which are subject to thermal cycling.

The American pressurised water and boiling water reactors also operate at relatively low temperatures. Maximum temperatures encountered in the primary circuit are in the region of 325 °C. Section III of the ASME Boiler and Pressure Vessel Code was specially developed for the BWRs and PWRs. This code, or derivations of it, is used extensively in most countries with a light water reactor programme.

Section III of the ASME Code is limited to 375 °C for ferritic materials and 425 °C for austenitics, and Code Case N47 extends Section III to higher temperatures. The impetus for this development was the liquid metal fast breeder reactor programme in the US. This code case is now used as the basis for design in the UK and other European countries. As its status as a code case implies, the design rules which it contains are not regarded as fully authenticated. A considerable body of work is underway in the US, under ASME and PVRC auspices, to remedy the more obvious shortcomings. Similar work in the UK is co-ordinated by the Fast Reactor Design Codes Working Group. The Codes and Standards Working Group of the European Commission's Fast Reactor Coordinating Committee provides a forum for the exchange of views within Europe.

The reason for the difficulty in establishing a fully comprehensive and fully validated set of design methods and design criteria is that the liquid metal fast breeder reactor, whether of European or US design, goes well beyond present experience. Maximum metal temperatures in the 550–575 °C region are relatively modest, and similar to those encountered in modern conventional power station practice. The novel features are the widespread use of austenitic steels in very large high temperature structures and the need to design against thermal shock and thermal fatigue. This is brought about by the very high power density of the reactor core, and the very good heat transfer properties of the liquid sodium. Reactor trips can lead to rapid changes of temperature in the primary circuit components; perturbations in sodium temperature can lead to high cycle thermal fatigue. Such problems are rarely met with in engineering design, outside the aero-engine industry where it is possible to test components, and indeed complete engines, for the full operating life.

The high capital cost of power plant, whether nuclear or fossil fuelled, means that economical production of electricity will only be achieved if plant can be designed to give high availability and good results for periods of upwards of 25 years. It is relatively easy to demonstrate that a reactor fully meets the very stringent safety requirements in regard to structural integrity when it first goes into service. What is needed is a formal procedure, which can be applied at the design stage, which gives reasonable assurance that the plant will continue to meet these safety criteria for the whole of its required life, without major outages and major repairs. Code Case N47 is, therefore, oriented towards long-term effects.

There are essentially two routes through N47. The designer can opt to use elastic stress analysis, and an associated set of failure criteria. Alternatively inelastic analysis and a different set of criteria can be used. In general, the elastic route is pessimistic in the sense that components which do not fully meet the criteria may very well be perfectly satisfactory for the full design life.

The inelastic route should inherently be less pessimistic, but has the disadvantage of increased complexity and an increased requirement for materials data. In general, 2-D and 3-D non-linear finite element calculations will be needed, which take into account time-dependent creep changes of stress and allow for cyclic loading. Such calculations are on the limits of the capability of present methods of computation. They will only produce meaningful results if adequate constitutive equations are available, which describe the behaviour of the materials under cyclic loading at high temperature. The calculated stresses and strains must finally be judged against suitable failure criteria to decide whether the design is adequate. As will be shown later, much further work is needed to define both constitutive equations and failure criteria before the inelastic route can be fully established.

3.1. Design in the Absence of Thermal Shock and Thermal Fatigue

Such considerations only apply to a limited number of LFMBR components, which are not subject to rapid changes of temperature, either under normal or abnormal conditions. At the same time, the majority of high temperature components in conventional power plant and in reactor systems, such as the AGR, are not subjected to rapid changes in temperature, and it is of interest to compare the N47 procedures with those normally applied for these types of plant.

The elastic analysis is essentially a modification of the traditional ASME Section III approach, in which the stresses are classified as primary membrane P_m, primary local P_L, primary bending P_B and secondary Q.

For high temperature design, the stress limits for levels A and B load-controlled quantities are specified as

$$P_m \leq S_{mt} \tag{5}$$

$$P_L + P_B \leq \begin{cases} 1{\cdot}5S_m \\ K_t S_t \end{cases} \tag{6}$$

where S_m and S_t are defined below, and K_t is a tabulated factor which allows for the creep relaxation of stresses in different geometries.

The omission of a limit on P_L and $P_B + Q$ is, at first sight, surprising. The intention appears to be that the requirements of Appendix T should also be applied to components that are not subjected to thermal shock and thermal fatigue. Tests 1 to 4 of Appendix T, to be discussed later, would, when applied to the total loading cycle, provide an assurance of shakedown similar to, and sometimes more restrictive than, that of inequality (4).

S_m is the tabulated design stress based on short-term yield, and S_t is the tabulated design stress based on creep properties. The time-dependent design stress, S_t, is taken as the lowest value of:

(a) The minimum stress to rupture, for the design life, divided by an appropriate factor.
(b) The stress which causes the material to enter tertiary creep in the required design life, divided by an appropriate factor.
(c) The stress to give 1% strain in the design life.

The values of S_m and S_t are tabulated for each of the materials used in LFMBR construction, and are also presented in graphical form. The design stress values differ in some respects to those used in other high temperature design codes. It is believed that such discrepancies, as exist, arise from the relatively short-term data, used by the N47 committee, which have been extrapolated to provide long-term design values.

The stress values quoted are not necessarily those which apply when irradiation effects are significant, or where the liquid sodium may affect the strength of the material. Here, the code case requires the designer to make any adjustments which are necessary.

Recent experience suggests that the high purity liquid sodium should have no deleterious effect on material properties. However, this may not be true of commercial quality sodium. Equally there may be problems with 'off-limit' sodium such as would occur following water ingress to the secondary circuit, or following leakage of lubricating oil from a pump. In the primary circuit it is quite clear that allowance has to be made for the effect of irradiation on long-term material properties when designing components adjacent to the core.

The elastic analysis route makes no specific reference to weld properties, although, in general, fillet welds are not permitted for pressure-retaining joints or for permanent structural attachments, and anything other than full penetration butt welds is discouraged. Attention is drawn in the code case to the potential for limited ductility of weld metal at elevated temperature and the potential for high stress concentrations in the heat affected zones of weldments. However, no guidance is provided on how the designer should undertake a quantitative assessment in order to follow the elastic analysis route.

In terms of creep rupture life, the elastic analysis route appears to be highly conservative. Applied to a boiler component, it would require much more local reinforcement in stress concentration regions than does, for example, BS 1113.

The inelastic analysis route provided in N47 is primarily concerned with the performance of structures under cyclic loading, as will be discussed later. As applied to structures under steady loading, the designer would be required to carry out time-dependent stress analyses, taking into account the creep of the material. The calculated maximum tensile principal strains at the end of design life would then be compared with the allowable values, which are:

(a)	Average through thickness	1·0 %
(b)	At surface, due to an equivalent linear distribution of strain through the thickness	2·0 %
(c)	Local strain at any point	5·0 %

Where welds are present in the component, the computation proceeds as if the structure were homogeneous using the parent material deformation properties. The maximum tensile principal strains calculated in this way are not permitted to exceed one half those given above, in the vicinity of a weld.

Welded transition joints between materials with dissimilar coefficients of expansion, for example between austenitic and ferritic steels, require special treatment. The designer is required to take into account the inelastic strains which can result in a significant localised strain accumulation near an abrupt change in mechanical properties.

Since, in this section, I am dealing with structures which are not subjected to large cyclic stresses, it would appear to be acceptable to use, as input to the inelastic calculations, deformation laws which are based on conventional constant load creep tests. However, to be compatible with the formalism of the code case, such deformation data would have to be consistent with the isochronous stress–strain curves which are given graphically in the code.

The use of maximum tensile principal strain as a design criterion is a departure from previous philosophy. The high temperature boiler and pressure vessel codes employ a stress-to-failure criterion, which implicitly assumes that, under multiaxial stress systems, a maximum principal stress failure criterion applies. This is supported, for the multiaxial stress systems encountered in pressure vessels, by test data. Considerably less information is available about how the strain to failure under multiaxial stress is related to the strain to failure in a conventional uniaxial creep rupture test. More work is needed to explore this before the strain criteria in N47 are fully validated, particularly in respect of materials which do not remain fully ductile throughout their design life, for example, because of thermal ageing or irradiation.

Experience with both good and bad welds in conventional boilers and pressure vessels shows that simply reducing the permitted strain in the weld regions will not provide an assurance that the plant will perform satisfactorily throughout its design life. Long-term tests on weld metal and complete weldments are necessary to demonstrate that the fabrication procedures are satisfactory. As far as transition joints between dissimilar metals are concerned, tests in the CEGB suggest that it is difficult to correlate lives with the stress analysis recommended by N47. Again, long-term tests on the actual design of joints are needed to demonstrate that they will perform satisfactorily.

3.2. Design Against Cyclic Loading at High Temperature

As pointed out earlier, many of the components in the LFMBR primary and secondary circuits are subjected to thermal shock and thermal fatigue. Code Case N47 represents a departure from all previously published design rules in that it provides a formalism to deal with these situations. This is described in Appendix T of the code case, which contains the rules for strain, deformation and fatigue limits at elevated temperatures.

Either elastic or inelastic analysis may be used. However, the code case points out that where creep effects are presumed significant, inelastic analysis is generally required to provide a quantitative assessment of deformation and strains. It states that elastic and simplified inelastic methods of analysis may sometimes be justified and used to establish conservative bounds of deformation and reduce the number of locations in a structure requiring a detailed inelastic analysis.

Two modes of failure have to be taken into consideration. Ratchetting may lead to unacceptably high deformation, so that the component ceases to function in an acceptable manner before the end of its design life. Alternatively, combined creep and fatigue may give rise to premature failure.

The starting point for the analysis is to define the load and temperature cycle to which the component will be subjected in service.

In the elastic analysis route, stresses in the component are calculated elastically at each point in the cycle. Parameters X and Y are then defined as follows

$$X = \left(P_{\mathrm{L}} + \frac{P_{\mathrm{b}}}{K_{\mathrm{t}}}\right)_{\max} \Big/ S_{\mathrm{y}} \tag{7}$$

$$Y = \frac{(Q_{\mathrm{R}})_{\max}}{S_{\mathrm{y}}} \tag{8}$$

where S_y is the average of the instantaneous yield values and the maximum/minimum average temperature during a cycle; $(P_L + P_b/K_t)_{max}$ is the maximum value of the primary stress, adjusted for bending via K_t, during the cycle; and $(Q_R)_{max}$ is the maximum range of secondary stress during the cycle.

Four tests are then applied. If any one of these is satisfied, deformations are deemed to be acceptable.

Tests 1 and 2 essentially ensure that stresses are always less than yield, and are thus very conservative. Shakedown will occur, so that there can be no progressive deformation. Test 4 caters for situations where the time spent at temperatures where creep takes place is an insignificant fraction of the total life of the plant.

Test 3 makes use of a diagram, originally due to Bree,[15] modified by O'Donnell and Porowski.[16] This diagram is reproduced in Fig. 3. The axes are in terms of the parameters X and Y, defined above. Shakedown is

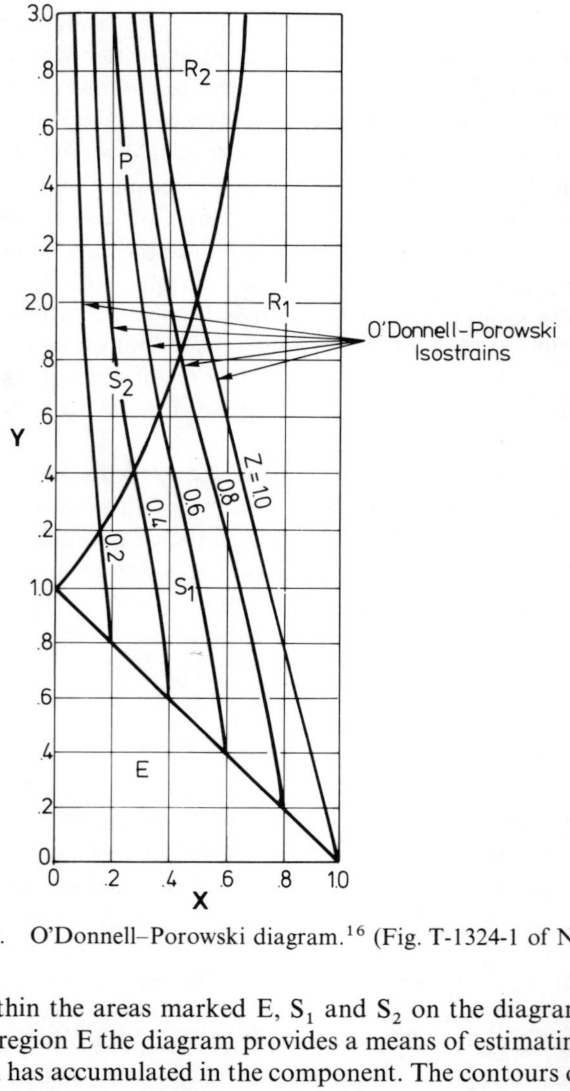

FIG. 3. O'Donnell–Porowski diagram.[16] (Fig. T-1324-1 of N47.)

achieved within the areas marked E, S_1 and S_2 on the diagram. Outside shakedown region E the diagram provides a means of estimating the total strain which has accumulated in the component. The contours of constant Z are used to define an effective creep stress σ_c, according to the equation

$$Z = \sigma_c / S_y \tag{9}$$

The creep ratchetting strain is determined by multiplying σ_c by 1·25, and reading off the appropriate isochronous stress–strain curve at this value.

In the elastic analysis route, combined creep and fatigue damage is evaluated by summing the fatigue and creep rupture damage, assessed separately, in a linear manner. A prerequisite for this form of creep-fatigue analysis is that one of Tests 1 to 4 has been satisfied with regard to ratchetting. The component is judged acceptable provided

$$\sum \frac{n}{N_d} + \sum \frac{\Delta t}{T_d} \leq 1 \tag{10}$$

where n is the number of times that the selected cycle is applied to the component; N_d is the total allowable number of cycles; Δt is the time duration of the selected load condition; and T_d is the allowable time duration at that stress level.

The fatigue damage is estimated by calculating the elastic strain range in the cycle, suitably modified to allow for stress concentrations due to local geometrical discontinuities and for plasticity.

The allowable number of cycles, N_d, is read off from one of two curves. T-1430-1 is used in conjunction with slow cycles and makes allowance for hold-time effects. T-1420-1 is used for continuous cycling, high cycle fatigue situations.

The inelastic analysis route presupposes that a computation has been carried out by taking into account both creep and plastic effects, to determine stresses and strains at all points throughout the loading cycle. The component is deemed to be satisfactory provided the strain limits given in the previous section have been satisfied.

Where a weld is present in the structure, the inelastic calculations are carried out, as discussed before, assuming the component to be homogeneous, and with the weldment having the deformation properties of the parent material. Inelastic strains in the weld region, calculated in this way, must not exceed one half of the specified strain limits for the parent material.

Estimation of creep and fatigue damage, as in the case of the elastic route, is carried out separately and the two forms of damage summed in a linear manner.

The equivalent strain range is obtained from the results of the inelastic computations. Using the continuous cycling curve T-1420-1 the allowable number of cycles, N_d, is obtained for that strain range. The fatigue damage $\sum n/N_d$ is then evaluated as before.

Creep rupture damage is estimated from the integral

$$\int_0^t \frac{dt}{T_d}$$

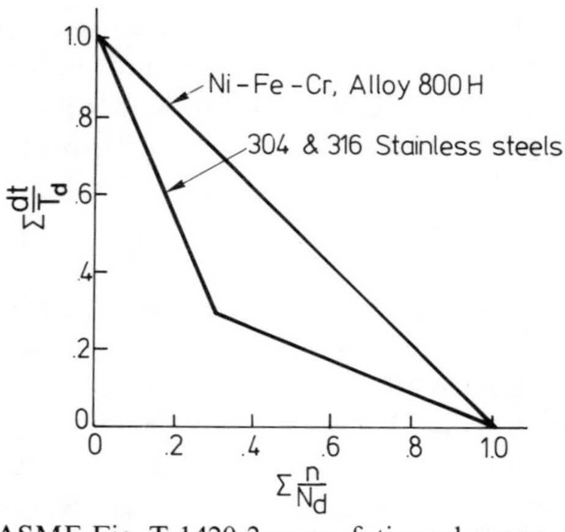

FIG. 4. ASME Fig. T-1420-2 creep–fatigue damage envelope.

where dt is the time spent at a certain stress and temperature, and T_d is the allowable time duration at that same stress and temperature permitted by the design stress values.

The total damage, D, is evaluated according to

$$\sum \frac{n}{N_d} + \int \frac{dt}{T_d} \leq D$$

For the component to be acceptable, the total damage, D, should not lie outside the envelope shown in Fig. T-1420-2 of the code, which is reproduced here as Fig. 4.

3.3. Further Research in Support of High Temperature Design Methods and Criteria for LFMBRs

Consideration of the underlying principles behind the code case indicates that a large on-going programme of materials research and structural integrity research is needed before fully validated methods and criteria are established. The very important question of large-scale tests on typical component features is outside the scope of the present paper, but deserves full discussion on another occasion. Although the greater part of the future research must relate to the highly important topic of components subjected to cyclic loading, some of it is equally applicable to outstanding issues relating to the non-cyclic situation.

First, there is a need for deformation data which describes the stress–strain–time behaviour of the material. Cyclic deformation behaviour is needed whether the elastic or inelastic route is followed.

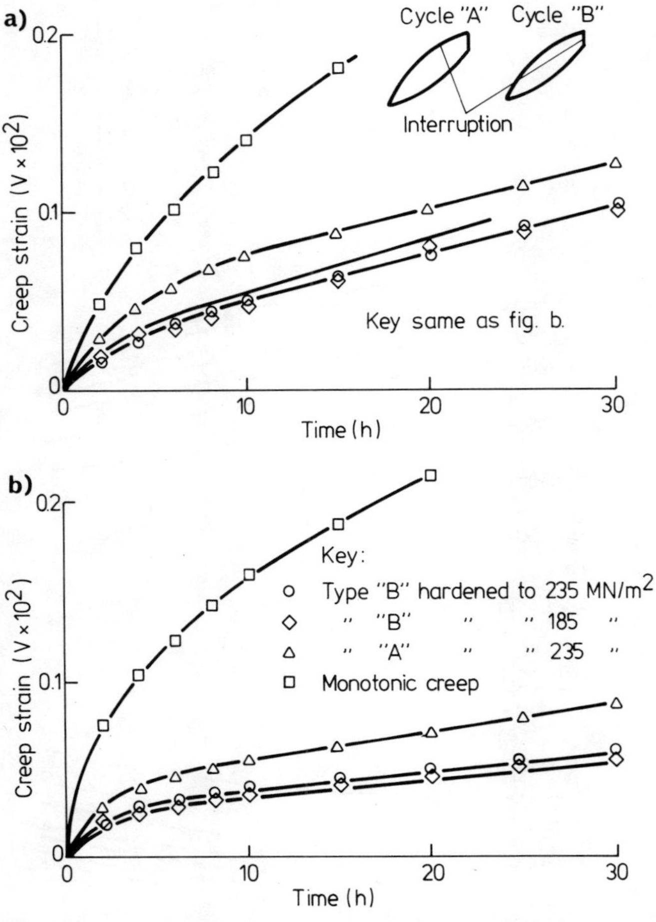

FIG. 5. Effect of prior cycling at 600 °C on subsequent creep strain. (a) Creep tests at 185 MN/m². (b) Creep tests at 154 MN/m².

The inelastic route implies that a computer analysis will be used to obtain the stresses and strains throughout an operating cycle. Sufficient work has already been completed to show that the cyclic deformation of the austenitic materials differs significantly from that observed in monotonic tests. Figures 5(a) and (b) and Fig. 6, taken from Ref. 17, will suffice to illustrate the point. For those who wish to pursue this topic further, additional information will be found in Refs 18, 19 and 20. It will be observed that cycling hardening takes place, so that the yield point is

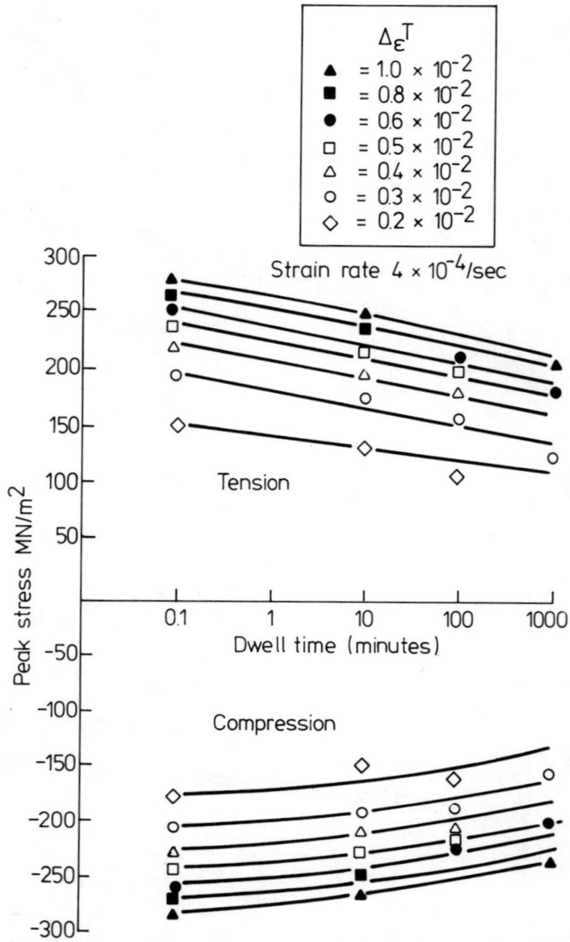

FIG. 6. Variation of peak stress with strain range and dwell times—cyclic tests at 600 °C.

effectively raised. This is to some extent off-set by hold periods at temperature, which effectively lead to some softening. The creep properties of the material are also affected by the cyclic plasticity, so that the stress relaxation which occurs during the hold period is markedly less than would be predicted from the conventional creep data. Without adequate constitutive equations to describe the cyclic behaviour of the material, the computation will be seriously in error, and so will the damage analysis which makes use of these computed stresses and strains.

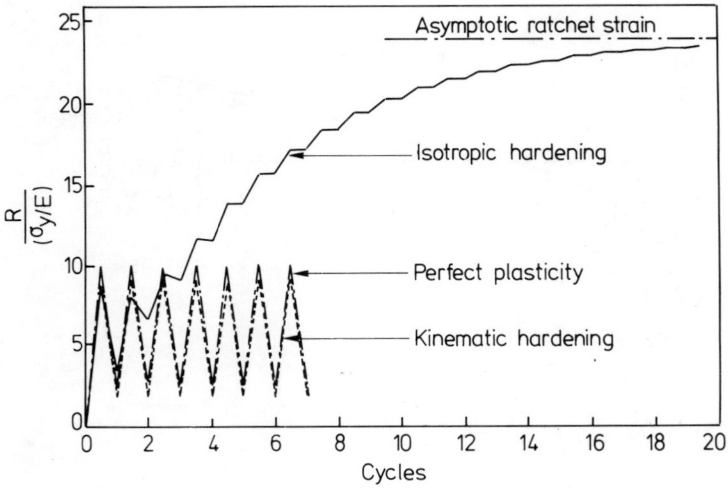

FIG. 7. Growth of ratchet strain for isotropic hardening in comparison with kinematic hardening and perfect plasticity solutions for an operating condition.

If the elastic analysis route is followed, the cyclic materials properties are of importance when the tests of Appendix T are applied. Not only do we need to know the increase or decrease in yield point which accompanies the cyclic loading, but we also need to know whether any hardening is isotropic or kinematic. This will affect the choice of S_y to be used in Tests 1 and 2, but by far the most important influence is on Test 3. Figure 7, taken from the work of Megahed[21] and Fig. 8 from the work of Goodman and Goodall[22] demonstrate the errors than can occur through the wrong choice of hardening model.

Much remains to be done. Development of a perfect constitutive equation, describing the complete behaviour of even one of the candidate materials, under all conditions of multiaxial cyclic stressing, would be an extremely expensive and time-consuming job. The aim must be to derive as simple as possible a description of material behaviour consistent with acceptable accuracy in the prediction of stresses and strains in a component. An important point to note is that there is no practical return for investigating phenomena which are only significant outside the operating range of temperatures, stresses and strains. (This is clearly an area where structural integrity and materials investigations must go hand in hand.)

Before leaving the topic of Test 3 of Appendix T, it should be noted that Fig. 3 of this Appendix has been derived from a theoretical investigation of

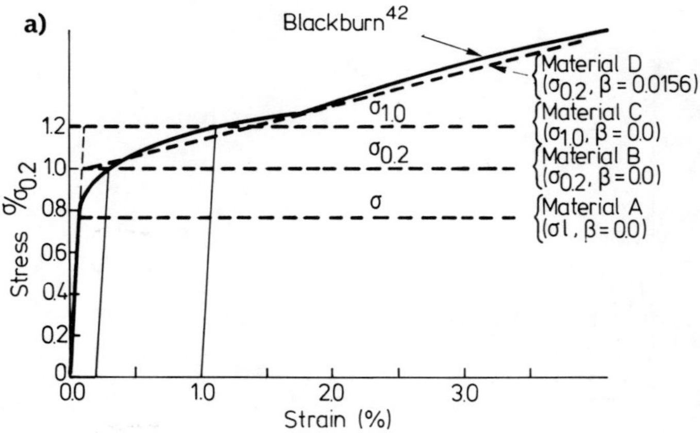

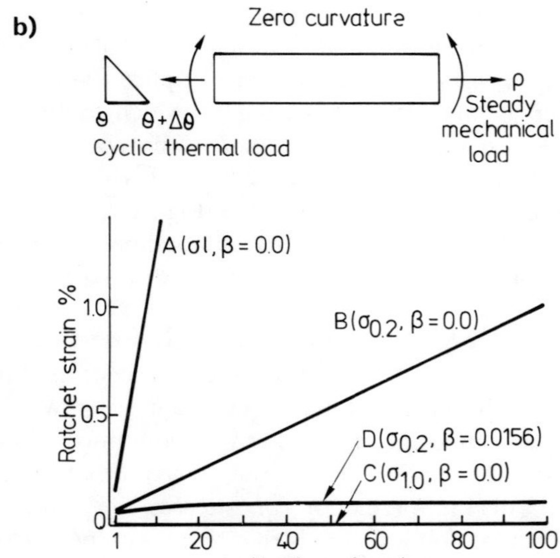

FIG. 8. Effect of different materials laws on calculated ratchet strain. (a) Various approximations to stress–strain curves for type 316 steel at 600 °C. (b) Calculated ratchet strains for Bree[15] beam materials A to D, as shown in Fig. 8(a).

single-sided thermal shock on a flat plate. Investigations of other geometries, subject to both single-sided and double-sided thermal shocks, and into tubular structures with an oscillating sodium level, show that each has a different shape of diagram. Reference to the publications of Goodman, Phillips and Ponter[23–26] shows that the differences are so significant that it would be inappropriate to use Fig. T-1324-1 for many types of component. Work is currently in hand, under the auspices of the CEC Codes and Standards Working Group, in conjunction with Professor Ponter at Leicester University, to provide a series of new diagrams, which will cover most of the geometries and loading systems commonly encountered in the LFMBR design.

Summation of creep and fatigue damage is another area where more research is needed. I will first deal with the high strain low cycle regime, usually associated with thermal shock. A large plastic strain is induced during a transient, and there then follows a period of operation when the induced stresses relax. Again, sufficient work has been completed to cast doubt on the validity of the procedures given in Appendix T and on the way in which design curves are derived from short-term tests.[27]

Materials data are conventionally obtained from short-term tests, accelerated by increasing strain ranges and reducing hold times. The laboratory tests are usually carried out with an idealised cycle, at constant temperature, in which the strain is held constant during the hold periods. Recent work by Hales[28] following a similar line of investigation to that pursued by Priest and Ellison[29] on ferritic materials has shown that, with all but the shortest hold times, there is no real fatigue–creep interaction. The damage mechanism is one of cavitation at the grain boundaries, such as would be apparent in a conventional creep test. However, the rate of initiation and growth of the cavities is enhanced by the cyclic loading. Hales[28] has shown that variation in strain rate round the cycle is an important factor, since the changes in ductility with strain rate have a profound influence on the accumulation of damage. What is now beginning to emerge is a sound fundamentally based method of extrapolating short-term laboratory data to operating conditions, to replace empirical procedures. At the same time, the work shows that many of the test results obtained under highly accelerated conditions cannot be extrapolated because a different damage mechanism occurs.

The low amplitude high cycle fatigue, which occurs in thermal cycling, also needs further investigation. A reasonable amount of data has now been accumulated on the high cycle endurance of candidate materials at the appropriate temperature. However, further work needs to be done to

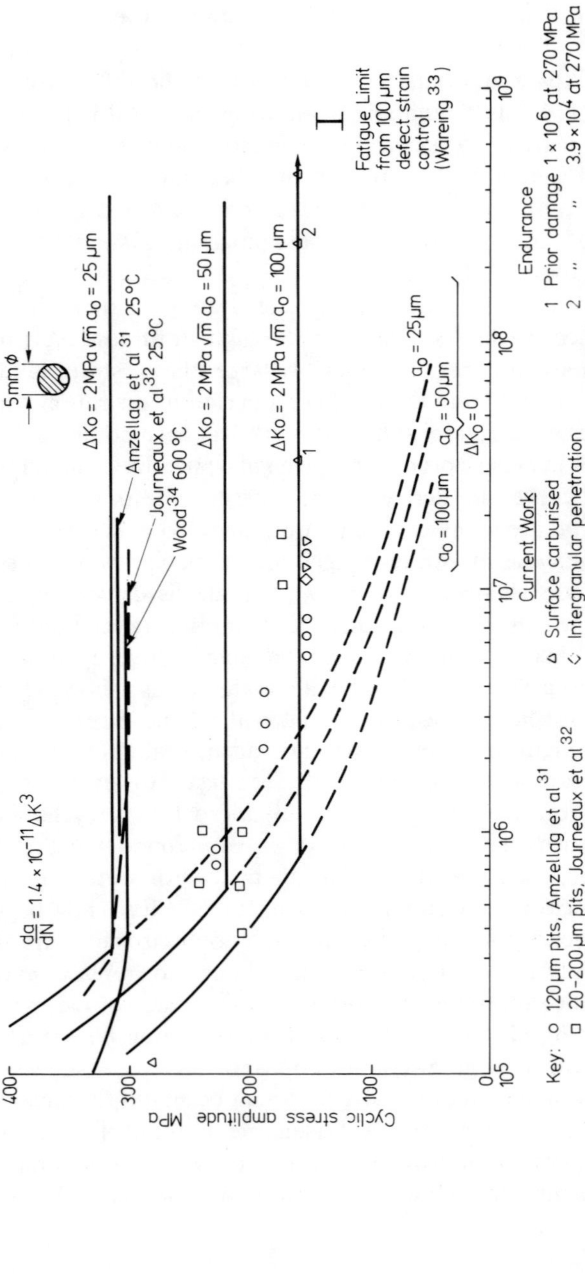

FIG. 9. Predicted and experimental fatigue curves.[30]

investigate the endurance of material which simulates the long-term ageing and irradiation that occur in service. Care is needed in using fatigue data from tests on conventional polished specimens. Figure 9, taken from Ref. 30, illustrates the reduction in high cycle fatigue life that can occur in type 316 steels when there are surface imperfections and scratches. An outstanding issue is the interaction of the damage produced in high cycle fatigue with the cavitation damage arising from thermal shock.

Although Code Case N47 provides no guidance on the assessment of components containing crack-like defects, it is quite clear that this will become an important issue. Although existing methods such as the CEGB R6 procedure[35] are available to demonstrate the safety of the highly ductile components when they go into service, much more work needs to be done before the integrity of the plant can be demonstrated after a long period in operation. The requirement is for the fracture assessment methods and the materials data to quantify crack propagation rates, under cyclic loading at high temperature, taking into account any degradation in materials properties which may have occurred due to thermal ageing, irradiation and occasional exposure to faulted sodium.

Another area where both structural integrity and metallurgical investigations are needed concerns the welds. Experience with both good and bad welds on conventional plant suggests that design lifetimes cannot be assured by the simple expedient of reducing the permitted strains by a factor of 2. Much more needs to be done to establish that whatever welds are selected they will not lead to a premature retirement of the reactor. Assurances are needed that the properties of the complete welded joint, not just the weld metal, will suffer no significant deterioration in service due to thermal ageing and to irradiation.

One of the most important requirements seems to be that the welds, and for that matter the parent material, should retain high ductility throughout their working life. The strain limits of N47 explicitly require that the material in the stress concentration regions must be able to sustain a 5% strain with a good margin of safety. In the stress concentration regions the stress system will be multiaxial, and this leads to speculation about the relationship between strain to failure and the multiaxial stress as compared with the elongation as normally measured in a conventional creep-rupture test. The work of Hales[28] demonstrates that ductility also plays a key role in determining the performance of the material under cyclic loading.

One important fact to emerge from recent research is that different casts of 316 stainless steel, all to the same nominal specification, can exhibit large differences in long-term ductility. The cause has been identified as a

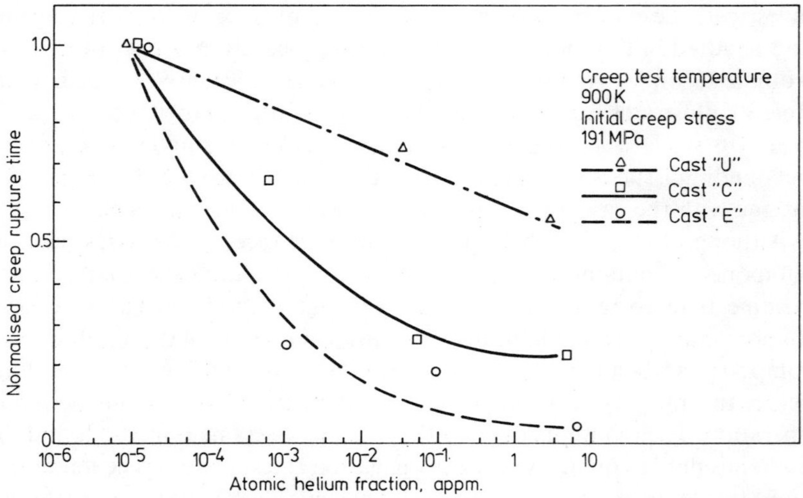

FIG. 10. Normalised creep–rupture times versus the atomic helium fraction (appm) for type 316 steel casts C, E and U after testing in irradiated conditions at 191 MPa and 900 K.

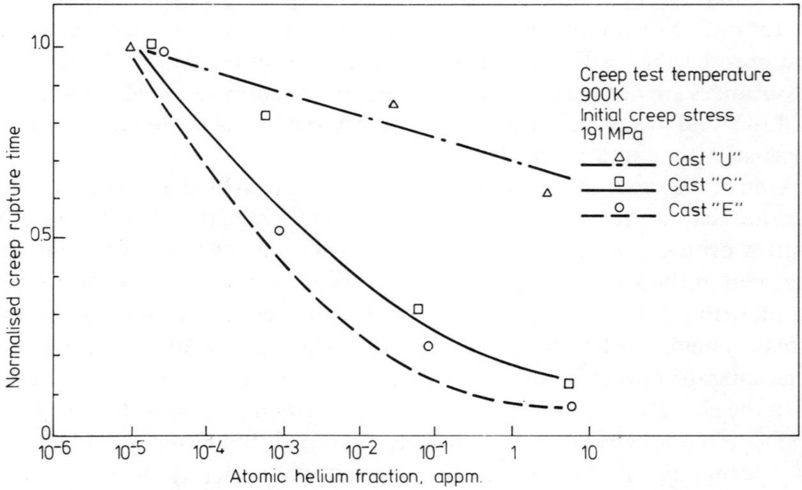

FIG. 11. Normalised creep ductility versus the atomic helium fraction (appm) for type 316 steel casts C, E and U after testing in irradiated conditions at 191 MPa and 900 K.

difference in the trace element quantities, and further information can be found in Refs 36 and 37. In general, boron up to 50 ppm is beneficial from the point of view of long-term rupture strength and rupture ductility, but steels with too high a boron content do not perform satisfactorily under irradiation. Figures 10 and 11, which illustrate this, are taken from Ref. 38. It is clear that a much tighter specification is required than is provided in national and international standards. Limits on boron in the range 20–50 ppm, and on carbon in the range 0·03–0·08 %, have been proposed by Wood.[39] It should be noted that, so far, investigations have almost exclusively been concerned with parent material. A parallel investigation is now required on weld metal and complete weldments.

Finally, there is scope for investigating alternative materials to replace the austenitic steels which have so far been favoured. Austenitic steels were originally chosen from considerations of creep-rupture strength at temperatures around 600 °C. However, because of the high coefficient of thermal expansion, high stresses are induced under thermal shock and thermal striping conditions. A slightly weaker material, with a lower coefficient of thermal expansion and with good long-term ductility and resistance to irradiation, could very well prove to be more suitable.

Consideration has been given to materials with reduced coefficients of expansion, such as alloy 718, for certain internal components, such as the above core structures. Unfortunately, detailed investigations showed that such materials were unsuitable. There are metallurgical problems arising from instability, and the materials may be notch-sensitive.[40]

More recently, attention has been turned to the 9Cr ferritic steels, which have a considerably lower coefficient of expansion than the austenitics. Preliminary tests show promise, and the strengthened versions appear to have adequate rupture properties and exhibit good ductility.[41] Tests under cyclic loading suggest good resistance to thermal shock. However, much further work needs to be done before they can be considered as a serious alternative to the austenitics. In particular, the long-term properties of weldments need to be investigated, cast-to-cast variations in the parent material need to be considered, and irradiation tests are required on both parent material and weldments.

4. CONCLUDING REMARKS

In drawing together the wide ranging observations that I have made about design matters, I would like to consider their implications in the context of the remaining sessions of this seminar. The subject has not been covered comprehensively, but I hope that I have drawn attention to the principal issues; no doubt they will be added to during the subsequent discussions.

As far as the design of fossil fired boiler plant is concerned, there seems little need for more metallurgical research beyond that which I have mentioned earlier. Unless new types of steel are introduced, there will be little need for further tests to obtain parent materials data, once the present programme of long-term creep testing is completed. A limited amount of research is justified to examine the long-term high temperature behaviour of existing types of weld and transition joint, with the object of eliminating any possible life-limiting features. The introduction of new welding processes and new consumables would also need high temperature materials tests to back them up.

The area for future development is related to the continuing fitness for purpose of plant which is reaching the end of its design life, and the associated question of remaining life prediction.

In the context of this paper, by far the largest research requirements are in the area covered by Code Case N47. This is not intended as a criticism of the code case as it appears at present, or those who are responsible for its production who have undoubtedly made very significant progress. Rather it is an acknowledgement of the difficulties inherent in providing unequivocal design rules in a region where there is little or no previous experience to provide a guide.

I have identified the topics where I believe further research is required, and I will not repeat them here. However, it is important to note that developments on structural integrity aspects must go hand in hand with developments on materials questions. Much of the research that is needed is at the interface between continuum mechanics and physical metallurgy, and success will only be achieved if the full forces of both disciplines can be brought together. Even the more straightforward questions of materials testing cannot proceed in isolation. Under the complex loading conditions, the engineer is unable to specify full materials data requirements until the preliminary results are available for stress analysis investigations.

I appreciate that the majority of the examples which I have discussed in the context of Code Case N47 have had a liquid metal fast reactor basis. This is because a large part of the world-wide research effort in the high temperature structural integrity field is associated with this type of reactor. However, many of the concepts now being explored have equal application in other areas of high temperature design. These are currently being used to examine life-limiting features in fossil fuelled boiler plant, and in high temperature reactors such as the AGR. They are beginning to be used in investigations into the thermal fatigue cracking of turbine valve chests and casings—components which, in the UK, are not designed according to strict code rules.

ACKNOWLEDGEMENT

This paper is published by permission of the Central Electricity Generating Board. It must be pointed out that the opinions expressed in it are those of the author, and do not necessarily represent a formal CEGB position.

REFERENCES

1. British Standards Institution (1983). BS 1113: 1983. *Draft standard specification for water tube steam generating plant.*
2. ASME (1981). *Boiler and Pressure Vessel Code Section III, Code Case N47-19*, The American Society of Mechanical Engineers, New York.
3. British Standards Institution (1982). BS 5500: 1982. *Specification for unfired fusion welded pressure vessels.*
4. British Standards Institution (1975). BS 806: 1975. *Specification for ferrous pipes and piping installations for and in connection with land boilers.*
5. British Standards Institution (1969). PD 6437: 1969. *A review of design methods given in the present standard codes and design proposals for nozzles and openings in pressure vessels.*
6. Leckie, F. A. and Ponter, A. R. S. (1972). *Archives of Mechanics*, **24**, 419–37.
7. Ponter, A. R. S. (1973). *J. Appl. Mech.*, **39**, 251–63.
8. Goodall, I. W. and Cockroft, R. D. H. (1973). *Int. J. Mech. Sci.*, **15**, 251–63.
9. Goodall, I. W., Cockroft, R. D. H. and Chubb, E. J., *Int. J. Mech. Sci.*, **17**, 351–60.
10. Brown, R. J., Cane, B. J., Parker, J. D. and Walters, D. J. (1981). *Creep and fracture of engineering materials and structures*, (eds B. Wilshire and S. Owen), Pineridge Press, Swansea, pp. 645–57.
11. Cane, B. J. and Manning, M. I. (1981). *Effect of test piece oxidation on stress rupture data*, CEGB Report LM/Mats/407.
12. British Standards Institution (1983). PD 6510: 1983. *A review of the present state of the art of assessing remanent life of pressure vessels and pressurised systems designed for high temperature service.*
13. British Standards Institution (1958). BS 1500: 1958. *Fusion welded pressure vessels for use in chemical, petroleum and allied industries. Part I: Carbon and low alloy steels.*
14. British Standards Institution (1965). BS 3915: 1965. *Specification for carbon and low alloy steel pressure vessels for primary circuits of nuclear reactors.*
15. Bree, J. (1967). *J. Strain Analysis*, **2**, 226–38.

16. O'Donnell, W. J. and Porowski, J. (1974). *WRC Bulletin*, **195**, 57–62.
17. Goodall, I. W., Hales, R. and Walters, D. J. (1981). *3rd IUTAM symposium on creep in structures, Leicester 1980*, Springer-Verlag, Berlin, pp. 103–27.
18. Goodall, I. W. (1981). *3rd int. seminar inelastic analysis in high temperature environment*, E. de F., Paris.
19. Pugh, C. E. and Robinson, D. N. (1978). *Nuclear Engineering and Design*, **51**, 269–76.
20. Turner, A. P. L. and Martin, T. J. (1980). *Met. Trans.*, **11A**, 475–81.
21. Megahed, M. M. (1981). *Int. J. Mech. Sci.*, **23**, 169–82.
22. Goodman, A. M. and Goodall, I. W. (1981). *Int. conf. mechanical behaviour and nuclear applications of stainless steel at elevated temperatures, Varese.*
23. Goodman, A. M. (1978). In: *Non-linear problems in stress analysis*, (ed. P. Stanley), Applied Science Publishers Ltd, London.
24. Phillips, J. (1982). *ASME J. Pressure Vessel Technology*, **104**, 232–8.
25. Goodman, A. M. (1981). *J. Engineering Structures*, **3**, 17–26.
26. Ponter, A. R. S. (1981). *Shakedown and ratchetting below the creep range, Parts I and II*, Leicester University Engineering Dept. Reports.
27. Coffin, L. F., Carden, A. E., Manson, S. S., Severnd, L. K. and Greenstreet, W. L. (1976). *Time dependent fatigue of structural alloys*, ORNL-5073, pp. 202–10.
28. Hales, R. (1980). *Fatigue of Eng. Struct and Matls*, **3**, 339–59.
29. Priest, R. H. and Ellison, E. G. (1981). *Materials Sci. and Engineering*, **49**, 7–17.
30. Marshall, P. (1983). *Conference on Designing with High Temperature Materials*, The Metals Society, Sheffield, to be published.
31. Amzellag, C., Rabbe, P., Lieurade, H. P. and Truchon, M. (1977). *Influence of Environment on Fatigue*, I. Mech.E, London.
32. Journeaux, G. E., Martin, J. W. and Talbot, D. E. G. (1979). *Mechanisms of Environment Sensitive Cracking of Materials*, Proc. Int. Conf., The Metals Society, Surrey (eds P. P. Swan, I. P. Ford and A. R. C. Westwood), p. 322.
33. Wareing, J. (1975). *Met. Trans.*, **8A**, 1367.
34. Wood, D. S. (1982). Unpublished work.
35. Harrison, R. P., Loosemore, K., Milne, I. and Dowling, A. R. (1980). *Assessment of the integrity of structures containing defects*, CEGB Report R/H/R6-Rev. 2.
36. Lai, J. K. L. and Wickens, A. (1979). *Acta Met.*, **27**, 217.
37. Bolton, C. J., Cordwell, J. E., Hooper, A. J., Marshall, P., Steeds, J. and Wickens, A. (1977). CEGB Report No. RD/B/N3991.
38. van der Schaaf, B. and Marshall, P. (1983). *Dimensional stability and mechanical behaviour of irradiated metals and alloys*, Vol. 1, BNES, London, pp. 143–9.
39. Wood, D. S. (1981). *Int. conf. mechanical behaviour and nuclear applications of stainless steels at elevated temperatures, Varese.*
40. Marshall, P. and Brinkman, C. R. (1981). *Nuclear Energy*, **20**(3), 257–69.
41. Patriarca, P. (1982). *A new Cr–Mo steel for commercial applications*, ORNL Technology Transfer Meeting.
42. Blackburn, L. D. (1972). *ASME Winter Annual Meeting*, New York.

2

High Temperature Component Design in the Chemical/Petrochemical Industry

Ph. G. Holl
Industrial Services, Kortenhoef, The Netherlands

1. INTRODUCTION

The objectives of a design code are:

(a) To establish essential basic principles for the guidance of the user, designer, fabricator, inspector and local authority.
(b) To indicate the selection of requirements to suit the specific service, hazards and demands of each application.
(c) To acquire the maximum balanced economical and safety requirements for design, materials, fabrication and inspection.

Design codes do not specify requirements for equipment which has been in service for its full lifetime.

The parts of equipment and piping in the steam generating and refinery process plants, operating under conditions where metallic components may deteriorate due to high temperature application, are discussed in this paper.

2. WHAT DESIGN CODES AND STANDARDS ARE USED?

For steam generating equipment ASME Code Section I or BS 1113 is used where no local code is in operation. For pressure vessels, ASME Code Section VIII Division 1 or 2 is used or BS 5500, with the additional requirements of the user. For the European countries, where each country has its own traditions, language, habits, design codes and laws, the use of BS 5500, supplemented with local requirements, is preferred.

The Tubular Exchanger Manufacturers Association (TEMA) code

covers the design, specification, construction and inspection of shell and tube heat exchangers. The TEMA standard is widely accepted, with the provision that 'ASME' is replaced by the local code.

The API recommended practice, API RP 530, is used for the calculation of heater-tube thickness.

For the design and construction of petroleum refinery piping ANSI B 31.3 is used, for flanges ANSI B 16.5 is used and for pipe fittings ANSI B 16.9 is used. For the construction, design and calculation of power piping and gas transmission systems ANSI B 31.1 and ANSI B 31.8, respectively, are applied.

Local pressure vessel design codes usually refer to the local material specification. Foreign material specifications may be used only with additional requirements. The ASTM specification is used as a 'lingua franca', and converted into an equivalent local material specification, which often leads to reductions in calculated wall thickness and costs.

Appendices A and B demonstrate the diversity of design codes, the differences in philosophy and legal status, and the difficulties inherent in these differences when interchanging code equipment.

3. PLANT DIFFERENTIATION

The manufacturing processes in the oil industry may be divided into three groups, viz. separation processes, conversion processes and treatment processes.

3.1. Separation Processes

The separation processes, such as distillation, absorption and solvent extraction, are based on the differences in the physical properties of the hydrocarbons. The design temperatures and design pressures are related to the boiling range of the feedstock.

The oil-fired heaters of the crude distillers and the high vacuum plants with average liquid temperatures of 350 °C and pressures of 3·5 bar gauge are the higher alloy metallic components, where corrosion of the inside of the heater tubes and transfer lines causes more mishaps than does temperature. A combination of the product velocity and the temperature may cause severe corrosion when naphthenic acid is present in the product (12 mm/year for carbon steel is possible). Sulphuric acid may cause a corrosion of 2 mm/year in carbon steel tubes at temperatures of about 400 °C. Faulty design of the heater or overloading may cause flame impingement of the tubes. Protection of the tubes may even increase this problem and cause tube burst.

3.2. Conversion Processes

The conversion processes involve a change in the structure of the hydrocarbons, either to improve the performance of the products or to convert one product into another. Plants involved are, for example, thermal and catalytic cracking, thermal and catalytic reforming, polymerisation, alkylation and isomerisation, of which only the cracking and reforming plant are of interest for this paper.

The thermal cracking units are working at temperatures of 520 °C and pressures of 25 bar. The conversion is brought about by heat alone. Wide experience has been gained with the Dubbs process for the application of different materials under higher temperatures and often under severe corrosive circumstances. Materials used were $2\frac{1}{2}$Cr–1Mo for the furnace tubes, 5Cr–$\frac{1}{2}$Mo for the vessels and 9Cr–1Mo for the pipelines. The coke formed in the heater tubes is removed regularly by burning out with controlled levels of steam and air. Excessive scale formation due to uncontrolled high temperatures during this de-coking process can necessitate the renewal of tubes.

In order to have a fast and rough indication of the heater-tube condition, hammer testing and measurement of the tubes are carried out whenever possible. An increase of the outside diameter of 3% is acceptable.

The fluid catalytic cracking plant consists of a reactor in which the cracking reaction takes place, the regenerator in which the formed coke is removed from the catalyst by combustion, and the stripper in which the oil products entrapped in the spent catalyst are removed by steam.

The design temperature of the regenerator is 650 °C at a pressure of 3 bar. The operating temperature is 640 °C at a pressure of 2·6 bar. The so-called cold wall design is obtained by the application of 12%Cr hexsteel with verilite and tuffmix filling. This cold wall design implies continuous surveillance of the wall with infra-red, visually or with temperature-indicating paint, in order to prevent severe wall damage occurring in the case where the concrete lining becomes damaged.

Pipelines transporting the catalyst from the regenerator to the reactor and from the reactor via the stripper back to the regenerator are protected against erosion by hexsteel and tuffmix. A typical example of erosion in a riser is given in Fig. 1. The compact design of the fluid catalytic cracker combined with the higher temperatures makes the use of flexible joints in the pipelines inevitable, with all their associated disadvantages.

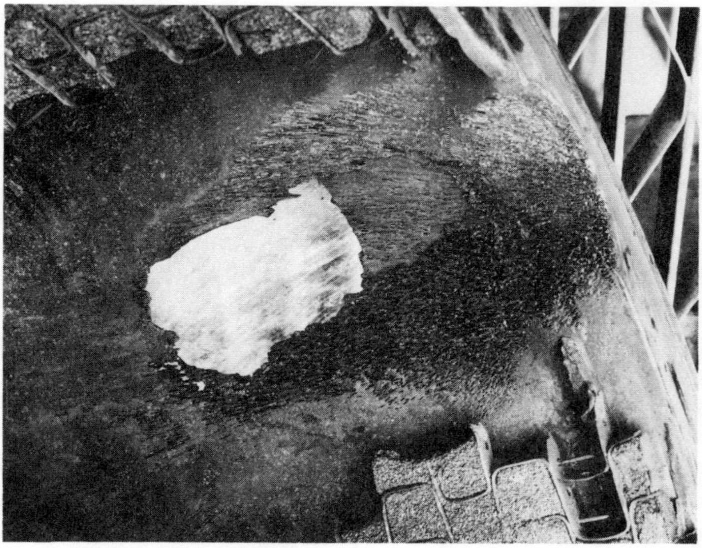

FIG. 1. Severe erosion in a riser (courtesy Shell).

In the thermal reforming plant, the intake is heated to the required reaction temperature of 550 °C in two stages. In the first cell (heating cell) the intake is heated up and in the second cell (soaking cell) the stock is maintained at temperature to allow time for the reaction to proceed. The operating pressure of 95 bar in the heater is reduced to 10 bar at the heater outlet and the temperature is then reduced by quenching oil injection. This option, using a complicated chemical rearrangement, can yield gasoline with a much higher anti-knock value.

A catalytic reforming plant, known as a platformer unit, makes use of a platinum-containing catalyst which improves the anti-knock quality of the gasoline. Components with low octane numbers are converted into components having high octane numbers, with hydrogen being liberated during the process. The unit consists of a prefractionator, three reactors, three heaters, an absorber and a product fractionator. Working temperatures in the reactors are about 450 °C; working pressures, depending upon the feedstock and the quality required, may vary from 15 to 50 bar. As hydrogen is present in this system, the materials for the piping, the pressure vessels, reactors and heaters are selected with the aid of Nelson curves. A low-alloy steel $2\frac{1}{4}$Cr–1Mo is used for the reactors, with a corrosion-resistant stainless steel cladding. Prevention of hydrogen attack of the steel components in hydrogenation plants is one of the major concerns in the process.

3.3. Treatment Processes

Treatment processes, where high temperature components are involved, include hydro-desulphuriser units in which sulphur components are removed by converting them into gaseous hydrogen sulphide by reaction with hydrogen in the presence of a catalyst. The feed mixed with hydrogen is heated in a furnace up to about 425 °C at a pressure of 40–60 bar, then passed through the reactor and led to the high pressure separator. The establishment of design criteria and the selection of materials to be used are identical to those of the platformer.

4. DESIGN OF HIGH TEMPERATURE EQUIPMENT

In the design code, the design pressure and design temperature are defined for a piece of equipment. No definitions are given for credible accidents and neither are guidelines for reliability evaluation of plant arrangements. The design code relates only to a piece of equipment and not to a combination thereof. Consequently, the margin between the average operation conditions and the design conditions can be far more than required by the applicable regulations. In practice, equipment designed for operation in the creep range on the basis of producing 1 % yield in 100 000 h will last much longer than it is designed for.

Lower stresses combined with lower operating temperatures than the design conditions specify may lengthen lifetime; other effects such as decarburisation of base metal, reheat cracking of deposit-welded lining, weld decay, etc. may influence material deterioration far more than the creep phenomena. Prediction of residual lifetime based on creep measurements remains uncertain until material investigation is undertaken. In view of this, design codes should state requirements for material investigations after design life operation. Regulations for the determination of possible material deterioration may imply that the policy of the inspection authority is to be changed.

Companies working with high temperature metallic components should be sufficiently aware of their responsibility for the safe and reliable working of their equipment. As governmental involvement in design, operation and maintenance is increasing all over the world, it is felt advisable to have detailed rules, prepared by both government and industry, in order to prevent unforeseen incidents.

5. WHERE IS EXTENSION OF DESIGN CODES REQUIRED AND WHAT RESEARCH WOULD HELP?

Design codes giving details of time-dependent design strengths should also state the minimum requirements for recording design life, starting with dimensional and microscopic investigations of the basic materials accompanied by pressure and temperature recording during operation, which could be replaced by a reliable on-the-spot non-destructive method for the assessment of material deterioration caused by creep, fatigue, reheat cracking, weld decay, etc. The influence of lower multiaxial stresses at higher temperatures on creep behaviour should also be investigated.

It is often difficult to convince inspection authorities not to repair detectable cracks in specific materials. The acceptability of crack opening displacement (COD) tests and the translation of the test results in lifetime assessment should be improved and introduced into the design code.

Developments in plant design increase the number of details not covered by the design code. The solution to pressure test equipment at 90 % of the theoretical yield with the application of strain gauges is felt to be insufficient. At the very least, a stress calculation with an accepted alternative, e.g. finite element method, should be carried out. Instructions for the application of such a method should be stated in the design code. For the information of shop and maintenance inspectors, the critical items of such a design could be stated on the design drawing.

Increases in operating temperature, pressure and throughput require an increase in wall thickness. Wall thicknesses of over 300 mm at temperatures of over 500 °C are already in use. The mechanical properties at elevated temperatures for these wall thicknesses are not covered by the material specification. This lack of information hampers the plant design. The design codes do not give restriction for wall thicknesses as long as the t/D ratio is within the scope of the code. Questions may arise if the existing codes, for the time-independent as well as for the time-dependent design strength calculations, are inconsistent with this thicker wall application.

The sensitivity of $2\frac{1}{4}$Cr–1Mo steel for temper embrittlement is a matter for concern and needs further investigation. The combination of creep and temper embrittlement for this materials application in hydrogenation plant is an uninvestigated area. Some design codes require notch toughness tests, such as the Charpy V notch test according to ISO, for material thicknesses over 50 mm and for low temperature application only. Through-thickness tests for heavier walls and for $2\frac{1}{4}$Cr–1 Mo steels after step-cool testing, could be introduced in the design code. Some steel manufacturers have introduced a factor expressing the susceptibility of $2\frac{1}{4}$Cr–1 Mo steel to temper embrittlement, based on Si, Mo, P and S contents; however, more metallurgical investigations should be carried out before introducing such a method into design codes or specifications.

APPENDIX A

Country	*Approval or accepted design code*	*Laws and/or regulations of boilers and pressure vessels*	*Approval organisation*	*Scope*	*Legal status*
Austria	No. 83 Dampfkesselverordnung, Gewährung von Ausnahmen von den Bestimmungen des Abschnittes IV. Gruppe II der Dampfkesselverordnung für Druckgas. No. 264 Werkstoff und Bauvorschriften für die Herstellung von Dampfkesseln	Verwaltungs-Entlassungsgesetz Bundesgesetzblatt No. 277 of 1925	Technischer Überwachungs-verein, Wien, Kruger Strasse 6, Wien 1	*Boilers* $t \geq 120$ °C $p \geq 1$ kgf/cm^2 $pv \geq 20$ *Vessels* $p \geq 0{\cdot}5$ kgf/cm^2 $pv \geq 300$	Compulsory by law
Australia	Standard Association of Australia Boiler Code AS CB1. BS 1515 and ASME I and VIII may also be accepted	Factory and machinery Acts of the various states	Respective State Statutory Authorities, e.g. Chief Inspector of Boilers and Pressure Vessels, Department Labour and Industry, 110 Exhibition Street, Melbourne 3000	Boilers and pressure vessels $p > 1$ kgf cm^2 $t \geq 99$ °C	Compulsory by law
Belgium	Code van Goede Praktijk voor de Constructie van Toestellen onder Druk N BN 121	Algemeen Reglement voor de Arbeidsbescherming	Gouverneur der Provincie	Boilers	Compulsory by law
Canada	Canadian Standards Association B 51, ASME Code	The boiler and pressure vessel acts of the various provinces	Chief Inspector of the province	Boilers and pressure vessels	Provincial regulation
Curaçao	Regels voor Toestellen onder Druk	Stoomwet Ned. Antillen N 163	Dienst Openbare Werken, Willemstad	Boilers steam raising equipment	Compulsory by law
Denmark	Swedisch (Tryckkörbsnormer), British Standards, ASME Code, German TRD and AD, Merkblätter	—	Direktoratet for Arbejdstilsynet, Upsalagade 20, Copenhagen Ø	Boilers and pressure vessels	Compulsory by law
Federal Republic of Germany	Technische Regeln für Dampfkessel AD Merkblätter Unfallverhütungsvorschriften	Various decrees of the Ministry of Labour	Technischer Uberwachungs-vereine (TUV)	Boilers and pressure vessels	Compulsory
France	Code de Construction des Appareils a Pression	Reglementation des Appareils a Pression de Gaz, Reglementation des Appareils a Vapeur	Ministere de l'Industrie, Direction des Mines, Service Technique, 97, Rue de Grenelle, Paris 7	Boilers and pressure vessels	Compulsory by law
German Democratic Republic	Vessels ABAO 850/1 Boilers ASAO 800–810	Werstoff und Bauvorsch-riften für Dampfkessel und Druckgefässer	Technische Uberwachung der DDR. Zentral Inspektion 108 Berlin, Leipzigerstrasse 5–7, German Democratic Republic	*Boilers* $p \geq 2$ kgf/cm^2 $HS \geq 0{\cdot}1$ m^2 *Vessels* $V \geq 0{\cdot}5$ litres $p \geq 0{\cdot}5$ kgf/cm^2	Compulsory by law
Italy	ANCC Code MS VSR and VSG collection	Regio Decroto Legge 9 Luglio 1926	ANCC, Via Urbana 167, 00184 Roma (and respective districts)	Boilers and pressure vessels	Compulsory by law

Material specification	*Basis of design*		*Hydrostatic test pressure*	*Approval of welders and welding procedure*	*Inspection during fabrication*
	Formulae for wall thickness of cylinders	*Max. obtainable joint efficiency*			
ONORM M3121	$\frac{PR}{SE} + 6\,\mathrm{mm}$	0·7	—	International certificate is acceptable	Interim and final inspection should be carried out by recognised organisation
HS 1548—1974	$t = \frac{PD}{zfn - P}$	1–	$Pd = 1{\cdot}5P \frac{fh}{fv} \frac{T}{T - C}$	AS CB14 parts I–IV SAA. Boiler Welders Certificate Code AS CB15 parts I–IV SAA Pipe Welding Code	Inspection by a recognised inspection authority
—	$e = \frac{PDX}{200ZRr} + 0{\cdot}5\,\mathrm{mm}$	0·9	$(1{\cdot}5P + 3{\cdot}5)\,\mathrm{kg/cm^2}$	No special rules	Inspection by a recognised inspection authority
ASTM	—	1	—	ASME Section IX	Inspection by Canadian National Inspection Board Committee
—	$d = \frac{PdD1}{2Zf1 - Pd}$	—	$1{\cdot}4Pd$ $1{\cdot}3Pd + 2$	Regels voor Toestellen onder Druk	Inspection by official authority
Nordic Group for Steel Regulations	—	—	—	Insta Welding Code	Inspection by a recognised inspection authority
DIN Normblatt	$s = \frac{DaP}{200(K/S)V + P} + C + C1 + C2$	1	$Pt = 1{\cdot}3Pd$	AD Merkblätt HP 2/1	Official inspection authority only
Norme Francaise Enregistree	$e = \frac{PD1}{2fZ - P}$	0·9	$Pe = 1{\cdot}3P - \frac{G_e20}{G_et}$ $\leq 1{\cdot}5P$	—	See: Article 4 de Decret 2 April 1926
			—	According to Code ABAO	Final inspection by TU Inspectors
UNI	$s = \frac{PDi}{200fZ - P}$	$Z = 1$	$Pd \leq 10\,\mathrm{kg/cm^2}$ $Pt = 1{\cdot}25Pd$ $Pd \geq 10\,\mathrm{kg/cm^2}$ $Pt = 1{\cdot}5Pd$	ANCC 'S' collection	ANCC inspectors only

APPENDIX A—*contd.*

Country	*Approval or accepted design code*	*Laws and/or regulations of boilers and pressure vessels*	*Approval organisation*	*Scope*	*Legal status*
India	Indian Boilers Act 1923	The Indian Boiler Regulation 1950	Chief Inspector of Boilers, Simla	Boilers and Pressure vessels	Compulsory by law
Japan	JIS. 8243—1968 Vessels JIS. 8201—1967 Boilers	Ministry of International Trade and Industry Ordinances	Ministry of Labour, 1-7-Okte-Machi, Chiyoda-Ku, Tokyo	*Boilers* $p > 1\,kgf/cm^2$ $HS > 1\,m^2$ *Vessels* $p > 1\,kgf/cm^2$	Compulsory by law
Netherlands	Rules for Pressure Vessels	Steamlaw and Steamdecree	Dienst voor het Stoomwezen, Stadhouderslaan Den Haag and District Authorities	Boiler and pressure vessels	Compulsory by law
New Zealand	New Zealand Boiler Code, New Zealand Fusion Welded Pressure Vessel Code	Boilers, Lifts and Cranes Act 1960	Marine Department, New Zealand	Boilers and pressure vessels	Mandatory
Norway	Forskrifter for Kjelr Tryckkorbsnormer, ASME Code, BS 1515, AD Merkblätter TRD Blätter	LOV AV 21 Mai 1971 Nr. 47 on Brannforlige Varer	Arbeidtilsynets Kjelkontroll, Box 8141, Oslo Dep.	Boilers and pressure vessels $P \geq 0{\cdot}5\,kgf/cm^2$	Mandatory
Singapore	British Standards	Factory Ordinance 1958	Ministry of Labour, Factory Inspectorate Section, Government Offices, Havelock Road	Boilers and pressure vessels	—
Sweden	Swedish Pressure Vessel Code	Arbitorshydslagrn: Svensk Forfattningssomling N1/1949	Swedish Steam Users Association	Boilers and pressure vessels	Mandatory
Switzerland	SVDB	Federal Regulations 9th April 1925 Verordnung Betreffend Aufstelling und Betrieb von Druckbehältern	Schweizerischer Verein von Dampfkesselbesitzern, Plattenstrasse 77, Zürich	*Boilers* $P \geq 0{\cdot}5\,kgf/cm^2$ $PV \geq 0{\cdot}2\,m^3$ *Pressure vessels* $P \geq 2\,kgf/cm^2$ $PV \geq 1$	Mandatory
South Africa	BS 1515, ASME Code, Rules for Pressure Vessels	Factories, Machinery and Building Work Act and Regulation	Divisional Inspector of Labour	—	For steam raising equipment compulsory by law
USA	ASME Code Section VIII Divisions 1 and 2 Section I Section II	Code of Federal Regulation No. 49 parts 1–199	National Board of Boiler and Pressure Vessel Inspectors	Boilers and pressure vessels	—
United Kingdom	BS 5500	Factories, Machinery Act	Recognised inspection authority	Boilers and pressure vessels	Required by insurance company

Material specification	Basis of design		Hydrostatic test pressure	Approval of welders and welding procedure	Inspection during fabrication
	Formulae for wall thickness of cylinders	Max. obtainable joint efficiency			
Indian Boiler Regulations Chapters II and V	$t = \frac{PR}{SE - 0{\cdot}58} + 0{\cdot}7\,\text{mm}$	0·9	$Pt = 1{\cdot}5Pd + 50\,\text{lbs}/d''$	By works inspector	By competent authority
JIS	$t = \frac{PRW}{200G\chi\eta - 0{\cdot}2P} + d$	1	$Pd \leq 1\,\text{kg/cm}^2$· $Pt = 2\,\text{kg/cm}^2$ $Pd \leq 4{\cdot}3\,\text{kg/cm}^2$, $Pt = 2\,\text{kg/cm}^2$ $Pd > 4{\cdot}3\,\text{kg/cm}^2$ $Pt = 1{\cdot}3 \times 4{\cdot}3 + 3\,\text{kg/cm}^2$	By Prefectural Labour Standard Bureau	Statutory inspectors
See rule 1 for pressure vessels	$d = \frac{PdD}{2Zf - Pd}$	1	$Pt = 1{\cdot}4Pd$ $1{\cdot}3Pd + 2$	By recognised inspector according to the rules	Intermediate inspection by competent inspector. Final inspection always by statutory inspector
Pressure Vessel Code	$\frac{PR}{SE} + 0{\cdot}06\,\text{in}$	0·8	$1{\cdot}5P$	Not required	Final inspection by surveyor of Marine Department
Nordic Group for Steel Regulations	—	0·9	$1{\cdot}5P$	Insta Recommendation	Competent inspection authority
British Standard steel	—	—	—	British Standard	—
Tryckkorbsnormer Nordic Steel Group	$S = \frac{DP}{200\frac{\delta}{SF}Z}$	1	$1{\cdot}3P$	Boiler Welding Code	Final inspection by SSUA
DIN	$\frac{PR}{SE} + 0{\cdot}04\,\text{in}$	1	$1{\cdot}5P$	Not required	Competent inspection authority
	—	—	—	According to code	Certificate of recognised inspection authority
ASTM	$t = \frac{PR}{2SE + 0{\cdot}4P}$	1	$1{\cdot}5P$	ASME Section IX	State recognised inspectors
British Standard steel	$t = \frac{PDi}{2fS - P}$	1	$Pt = 1{\cdot}3\left(P + \frac{fn}{fi}\frac{T}{T - C}\right)$	According to Code	Recognised inspectors

APPENDIX B: PRESSURE VESSELS

GUIDE TO ASME SECTION VIII, DIVISION 1[a]

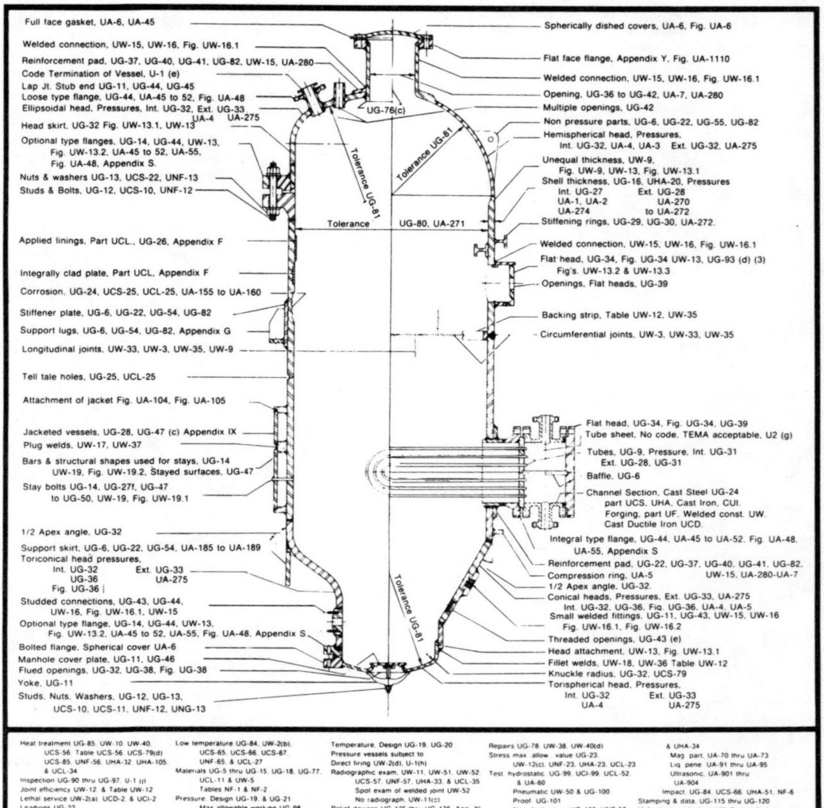

[a] Courtesy Missouri Boiler and Tank Co., Division of Nooter Corp., St Louis, Mo., USA.

GUIDE TO BRITISH CODE BS 5500[a]

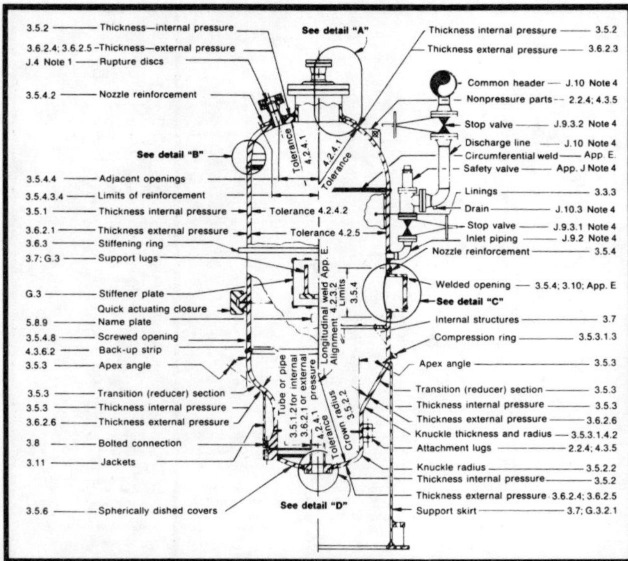

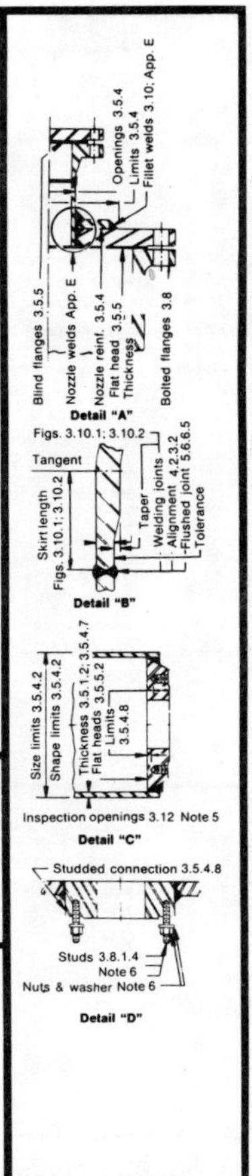

Definitions	1.4
Design pressure	3.2.3
Design temperature	3.2.4
Maximum allowable pressure	Note 2
Corrosion allowance	3.3
Joint efficiency	Note 3
Heat treatment	4.4
Materials	Section Two
Stamping	5.8.9
Loadings	3.2.1
General tests	Section Five
Hydrostatic test	5.8.1; 5.8.2; 5.8.3; 5.8.5
Pneumatic test	5.8.1; 5.8.2; 5.8.4; 5.8.5
Proof test	5.8.6
Low temperature service	3.2.4; App. D
Non-destructive testing	5.6
Inspection	Section Five
Repairs	4.4.3.4; 5.7.4
Construction category	3.4.1 & Table 3.4

NOTES

1. Refer also BS.2915: Bursting discs and bursting disc assemblies.
2. Not defined. Maximum allowable working pressure is equal to design pressure as calculated using actual metal thicknesses - less corrosion allowance - provided.
3. BS.5500 does not use the concept of a "weld joint efficiency factor".
4. Refer also BS.0000: Safety valves for use in the chemical, petroleum and allied industries (in course of preparation as at Dec. 1978).
5. Refer also BS.470: Access and inspection openings for pressure vessels.
6. Refer also BS.4882: Bolting for flanges and pressure containing purposes.

Annual up-dating supplements and code case interpretations are supplied automatically when so specified at time of purchase of basic BS.5500.

BS.5500 replaces BS.1500 Part 1 and BS.1515 Parts 1 and 2 which have been withdrawn. BS.1500 Part 3 will be withdrawn in 1979 following the publication of supplementary material to BS.5500 covering aluminium and aluminium alloys.

GENERAL REFERENCES

Heat exchanger tubesheets	3.9
Tubeplates (waste heat boiler type)	3.5.5.3
Stayed surfaces	3.5.5.3
Design criteria for loadings and components not explicitly covered by design formula	Appendix A
Shells under combined loadings including wind and earthquake	Appendix B
Assessment of vessels subject to fatigue	Appendix C
Calculation of stresses from local loads, thermal gradients etc.	Appendix G
Derivation of material nominal design strengths	Appendix K
Welding procedure qualification tests (see also BS.4870: Part 1)	5.2
Welder and operator qualification tests (see also BS.4871: Part 1)	5.3
Production control weld tests	5.4
Welding consumables	4.3.2
Vessels having finite design life (creep or fatigue conditions)	3.2.4

[a] J. W. Strawson, Oil Companies Materials Association, London, UK.

GUIDE TO A. D. MERKBLATT CODE[a]

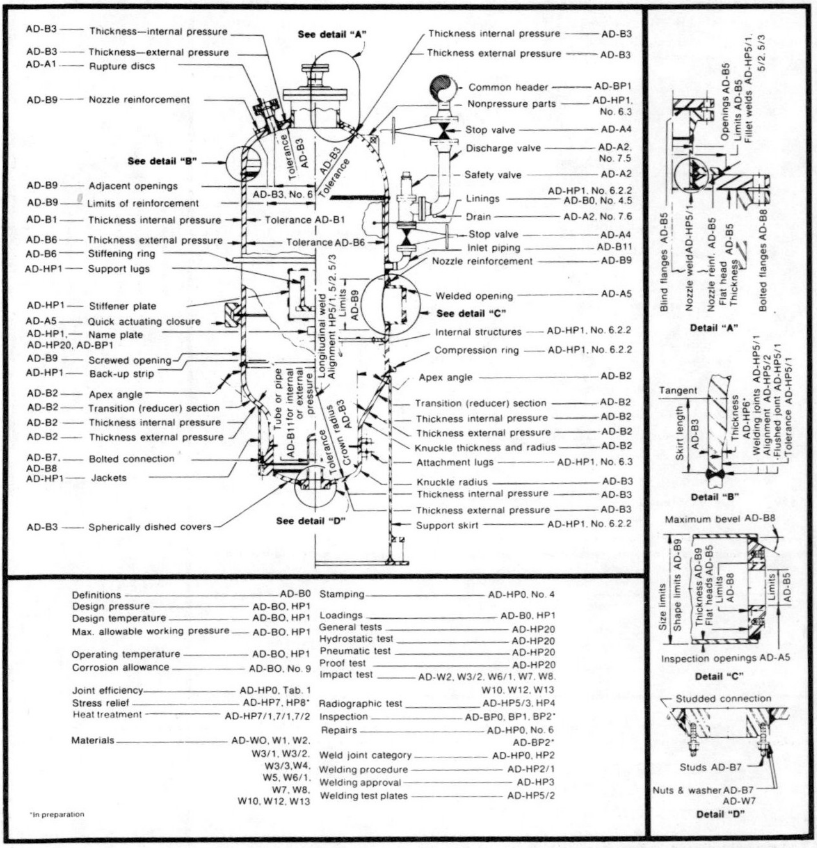

[a] H. Steffen, Vereinigung der Technischen Uberwachungs-Vereine e.V., Essen, West Germany.

GUIDE TO DUTCH STOOMWEZEN CODE[a]

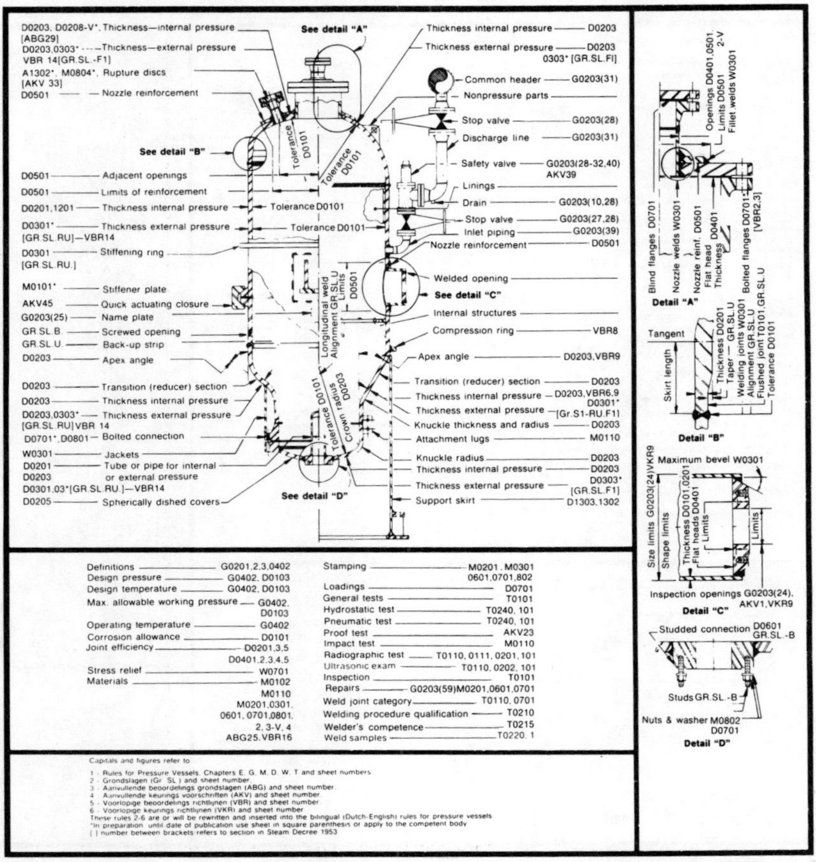

[a] P. van Rossen, Dienst voor het Stoomwezen, 's-Gravenhage, The Netherlands.

GUIDE TO SWEDISH PRESSURE VESSEL CODE[a]

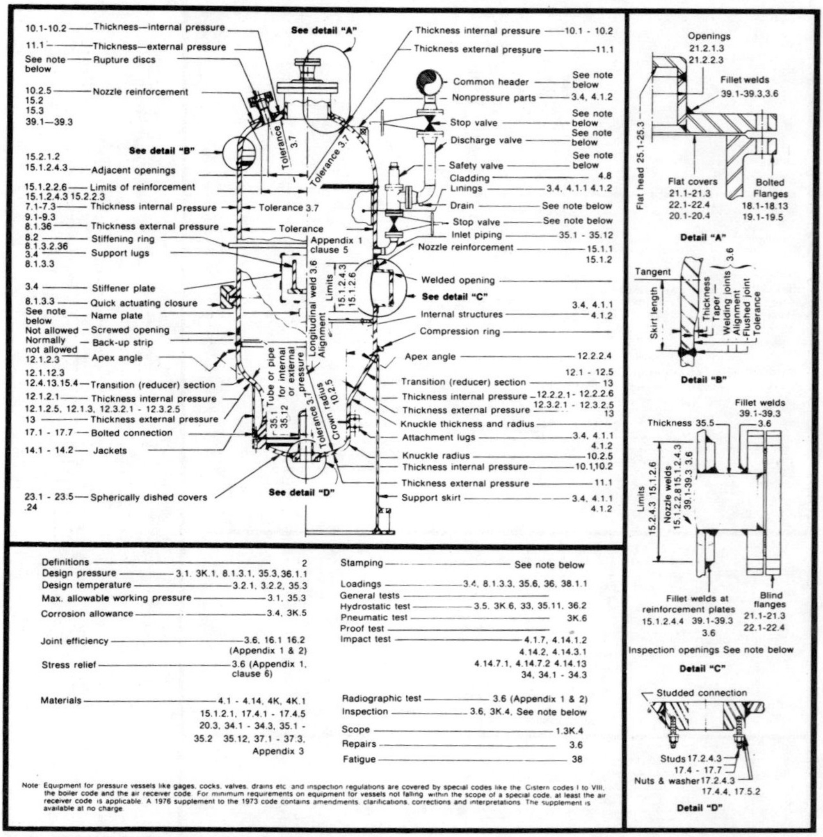

[a] I. Berglund, Tryckkärlskommissionen, The Swedish Pressure Vessel Commission, Stockholm, Sweden.

GUIDE TO ITALIAN PRESSURE VESSEL CODE[a]

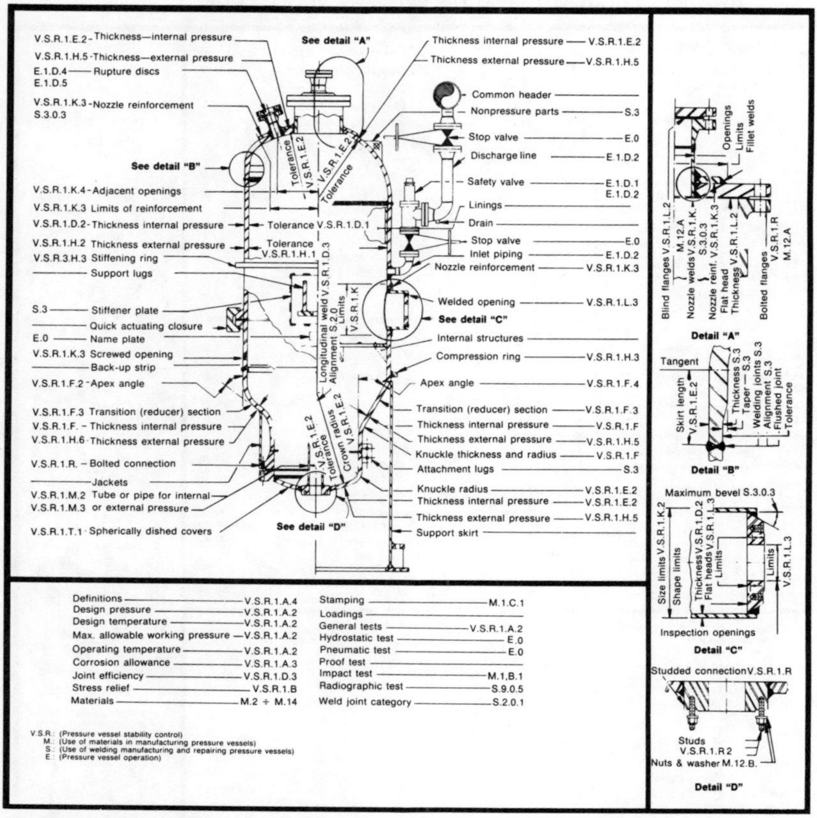

[a] R. Mazzoncini, Brescia, Italy.

GUIDE TO JAPANESE PRESSURE VESSEL CODE[a]

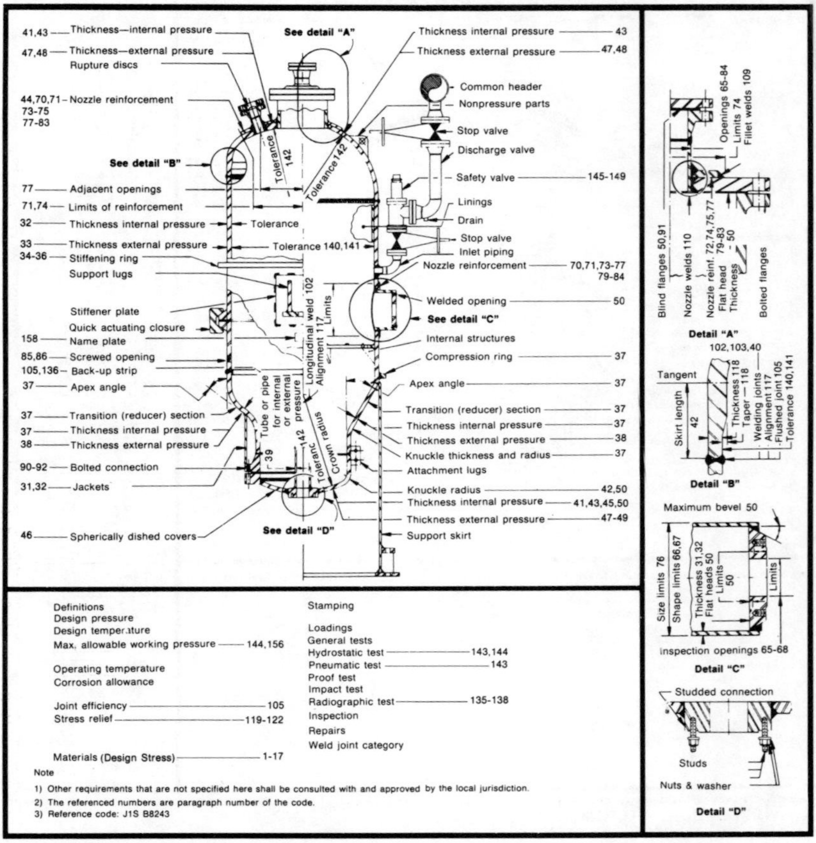

[a] M. Koike, Nippon Kokan K.K., Tokyo, Japan.

GUIDE TO JAPANESE STANDARD PRESSURE VESSEL CONSTRUCTION[a]

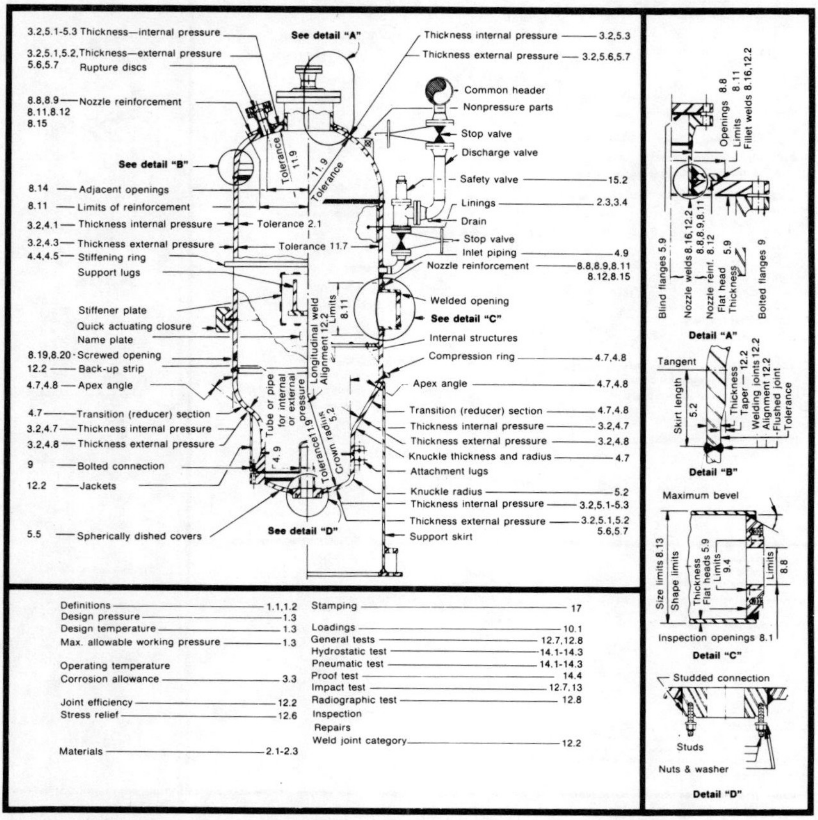

[a] M. Koike, Nippon Kokan K.K., Tokyo, Japan.

GUIDE TO JAPANESE HIGH PRESSURE GAS CONTROL LAW[a]

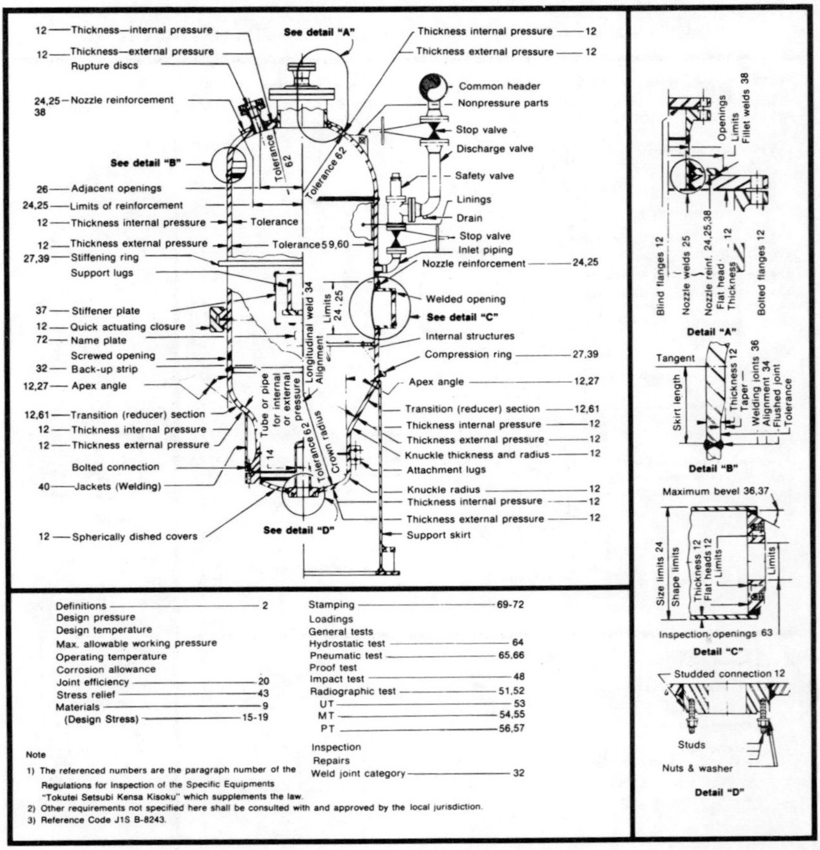

[a] M. Koike, Nippon Kokan K.K., Tokyo, Japan.

3

Acquisition, Analysis and Interpretation of Creep and Creep-Rupture Data

R. W. EVANS and B. WILSHIRE

Department of Metallurgy and Materials Technology, University College of Swansea, UK

1. INTRODUCTION

When a stress is applied to a metal or alloy, after the virtually instantaneous strain which occurs on loading, the material may continue to deform in a time-dependent manner and may eventually fail. This time-dependent deformation, known as creep, can occur at all temperatures above absolute zero. However, the precise mechanism by which creep and creep fracture take place and the detailed strain/time behaviour exhibited depend not only on the material selected and its precise metallurgical state but also on the applied stress and temperature conditions experienced.

The technical relevance of creep and creep fracture became apparent with the rise in the operating temperatures employed in electricity generating plant, petrochemical installations and aero-engines. In turn, these developments focussed attention on the need for improved materials having a satisfactory combination of high temperature properties and resistance to degradation in order to withstand the arduous and often hostile service environments imposed for periods up to, and sometimes beyond, the design lives anticipated. The metallurgical complexity of most commercial high temperature materials has meant that studies aimed at identifying the mechanisms by which creep and creep fracture occur have usually been undertaken using pure metals and relatively simple alloys. Similarly, most fundamental investigations and even the majority of engineering design procedures are based on uniaxial test data whereas, in practice, components and structures are generally required to operate under complex non-steady stress and temperature conditions. In the present work, an appraisal has been made of uniaxial test procedures and the approaches widely used to explain uniaxial creep behaviour in order to provide a basis for assessment of new concepts for the acquisition and interpretation of the high temperature materials data essential for engineering design.

2. TENSILE CREEP TESTING

National and international standards cover the methods which should be adopted in order to acquire the proof stress and uniaxial creep data needed for materials assessment and for design of components for high temperature service (e.g. BS 3688 and BS 3500). Procedures are then specified for short-term creep tests up to about 100 h duration which are normally undertaken for quality assurance of the relevant materials and also for the long-term tests used to obtain design data. A further subdivision is then usually made depending on whether stress-rupture or creep properties are required. With stress-rupture tests, the experimental procedures are relatively simple since measurements are made only of the time to fracture, t_f, and the creep ductility, ε_f (which can be determined from the specimen dimensions before and after testing). More complex test equipment able to monitor the time-dependent changes in specimen dimensions must then be employed to obtain creep data, such as the time to attain a strain of, say, 1%.

2.1. Constant-Load and Constant-Stress Equipment

The type of test most frequently used for determination of the creep and/or stress-rupture properties of metals and alloys involves the application of a constant load to a specimen maintained at a constant temperature. When a tensile load P is applied to a testpiece of initial cross-sectional area A_0, the total strain recorded after a time t is made up of the virtually instantaneous strain on loading and a time-dependent strain. The resulting nominal strain is defined as $(\ell - \ell_0)/\ell_0$, where ℓ_0 and ℓ are the gauge lengths recorded before loading and after a time t. With this 'constant-load' procedure it is essential to recognise that the creep and/or the stress-rupture properties are being considered in relation to the initial specimen dimensions and that the stress quoted is always the initial stress (P/A_0) applied to the specimen.

Many different types of constant-load equipment are available. In general, except for wire and strip specimens, the dead loads required with direct loading are too large for convenience and thus a lever arm is introduced into the design (Fig. 1). With a simple lever, although the tensile

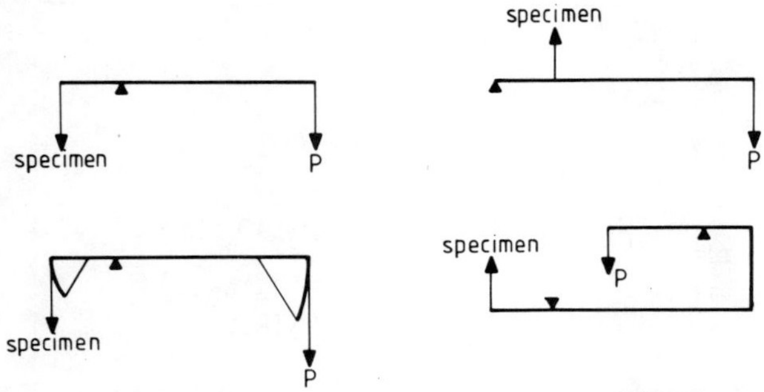

FIG. 1. Various lever configurations for tensile creep machines.

load can be maintained constant as the testpiece extends, the true stress on the specimen increases as the cross-sectional area decreases with increasing strain. If only very low total strains are involved, little error is caused by assuming that deformation is taking place under a constant stress (P/A_0). However, many materials can exhibit overall strains in the range 10–40 % so that the actual stress on the specimen can increase considerably throughout the test. The 'true stress', σ, at any instant is given by

$$\sigma = \frac{P}{A_0} \exp(\varepsilon)$$

where ε is the true strain at any time which is defined as

$$\varepsilon = \int_{\ell_0}^{\ell} \frac{d\ell}{\ell} = \ln \frac{\ell}{\ell_0}$$

The problems associated with 'constant-load' testing of ductile materials can be illustrated most easily by reference to a hypothetical situation in which the true creep strain increases linearly with time until fracture occurs at a strain of, say, 0·1 under 'constant-stress' conditions, i.e. the stress, σ, rather than the load, P, remains constant throughout the test. Using constant-load equipment, instead of a simple linear strain/time relationship, the strain rate would appear to increase continuously throughout the test and the rupture life would be considerably shorter than that for constant-stress conditions (Fig. 2).

Constant-stress conditions can be created by systems which automatically reduce the applied tensile load in proportion to the decrease in specimen cross-section (which, assuming constant specimen volume, is equivalent to the increase in gauge length). This may be achieved in several ways, e.g. by special spring loadings, by selected geometrical linkages, by

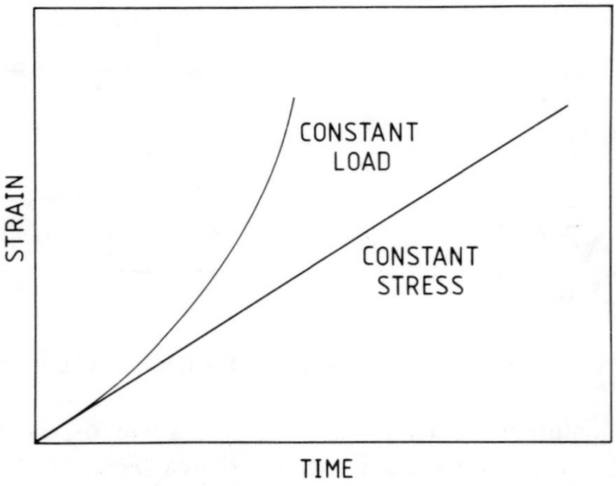

FIG. 2. Schematic illustration of the difference in the strain/time behaviour expected for constant-stress and constant-load testing. Assuming that a constant creep rate occurred throughout a constant-stress test, the creep curve would appear as a continuous pseudo-tertiary curve under constant-load conditions due to the gradual increase in stress as the cross-sectional area of the testpiece decreases progressively with increasing strain.

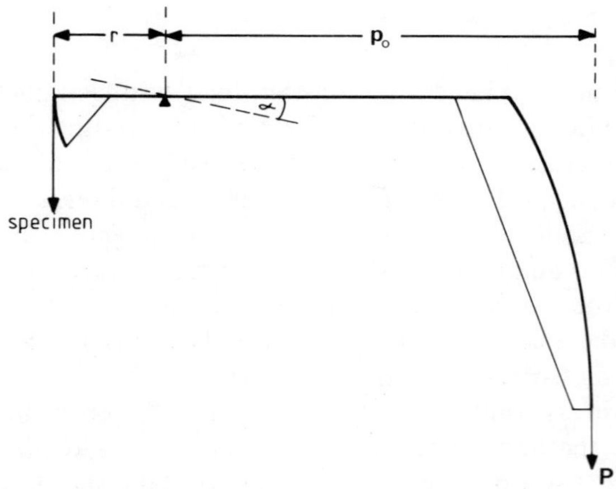

FIG. 3. Schematic illustration of a profiled cam arrangement for constant-stress testing.

employing servo-feedback systems for load control, or even by the use of shaped loading weights which gradually descend into a liquid so that the Archimedian upthrust reduces the load progressively as strain accumulates. However, the most commonly adopted method involves the incorporation of a profiled cam in the loading arm. One such system is illustrated in Fig. 3. In this case, a semi-circular cam of radius r is located on the specimen side which maintains a constant distance from the fulcrum to the specimen loading train. The actual geometry of the profiled cam situated on the loading side can be calculated as follows. Let A_0 and ℓ_0 represent the original cross-section and initial gauge length, respectively, while A and ℓ define these quantities after a time t. If p and r are the horizontal distances from the fulcrum to the loading line and the specimen assembly (Fig. 3), then the initial stress with the lever arm horizontal becomes

$$\sigma = \frac{P}{A_0}\frac{p_0}{r}$$

During the time t let the specimen extend by an amount which causes rotation of the lever arm through an angle α so that

$$A = \frac{A_0 \ell_0}{(\ell_0 + r\alpha)}$$

where $r\alpha$ is the increase in gauge length. In order that the stress remains constant

$$\frac{P}{r}\frac{p_0}{A_0} = \frac{Pp(\ell_0 + r\alpha)}{r_0 \ell_0 A_0}$$

or

$$p = p_0 \ell_0/(\ell_0 + r\alpha)$$

Machining a cam to follow this expression relating p to α adjusts the lever ratio P/r so that the stress σ remains the same with increasing strain. However, since this expression determining the cam profile contains the term ℓ_0, different cam geometries are usually required for different initial specimen gauge lengths.

2.2. Creep Strain Monitoring Systems

Although special configurations must be developed for the testing of wire or strip, most creep or stress-rupture studies are performed using

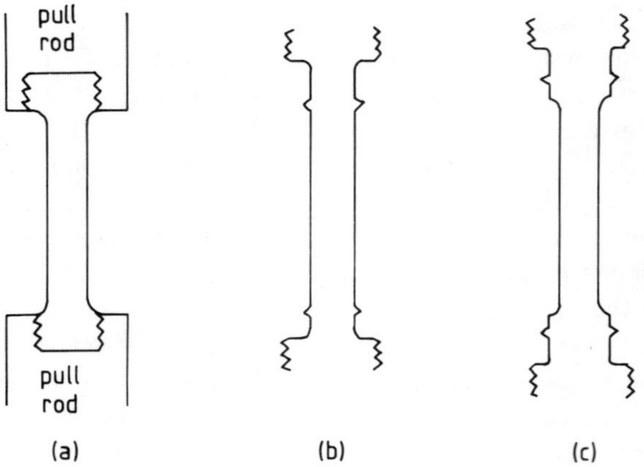

FIG. 4. Diagrammatic representation of: (a) specimen gripping arrangement; (b) ridges on gauge length of tensile specimen; and (c) ridges in shoulders of a tensile creep specimen.

cylindrical specimens which are referred to as 'proportional' testpieces when

$$\ell_0 = 5{\cdot}65\sqrt{A_0}$$

or when $\ell_0 \simeq 5d_0$, where d_0 is the initial diameter of the gauge length. Specimens are usually produced with threaded ends which screw into pull bars made out of materials such as stainless steel or nickel-base superalloys. These types of material are normally adequate for temperature ranges up to ~ 1250 K. In order to minimise bending stresses, the specimen gauge length must be coaxial with the threaded ends, the specimen and pull rods must be accurately aligned, and universal joints with coplanar knife edges should be incorporated into the loading train (Fig. 4(a)).

The successful design of creep, as opposed to stress-rupture, equipment also requires a satisfactory method whereby the changes in gauge length with time can be measured. There is, as yet, no wholly satisfactory method of realising this aim. The most widely accepted method of achieving accuracy involves machining of ridges onto the specimen gauge length (Fig. 4(b)) but this solution does modify the stress distribution near to the ridges. An alternative approach is to machine the ridges onto shoulders between the gauge length and the threaded ends (Fig. 4(c)) but it then becomes necessary to derive an 'apparent gauge length' over which the creep strain accumulates.

An extensometer, with arms leading out of the furnace, can be clamped

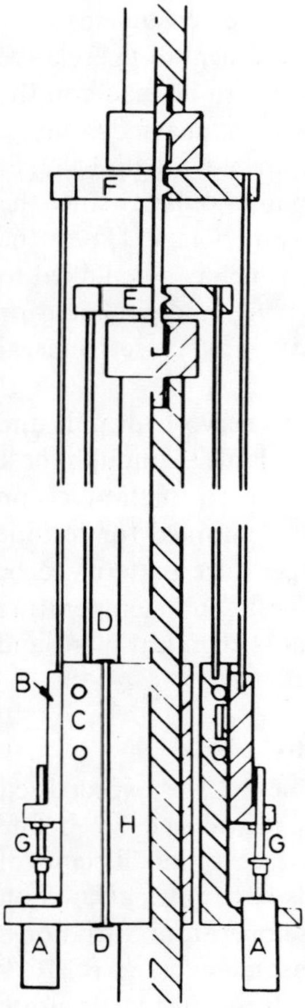

FIG. 5. Specimen gripping arrangement and extensometer assembly. In this case, split grips E and F clamp onto ridges on the specimen gauge length. Magnetic forces between the extensometer ends B and C ensure that relative movement of each pair of extensometer arms occurs only in the vertical direction. Vertical movement of block C with respect to the lower pull rod H is achieved by beryllium–copper leaf springs D. Relative movement of the extensometer arms is measured by differential capacitance transducers, A. Readjustments for large extensions can be made by means of the micrometers, G.[1]

onto the specimen ridges which allows the change in gauge length between the ridges to be measured from the relative movement of the extensometer arms. Clearly, in order that this relative movement does represent accurately the actual change in gauge length, the extensometer arms must be carefully aligned and must remain parallel throughout the test. The complexity of the extensometer system and the type of strain-measuring device used to monitor the relative movement of the arms depend on the accuracy with which the creep strains must be measured. By way of illustration, a stable strain-monitoring system which has proved particularly suitable for high-precision strain recording in tests carried out using constant-stress equipment is outlined in Fig. 5.[1] With this system, changes in gauge length are monitored by means of two pairs of extensometer arms (each pair on opposite sides of the specimen). Measurement of the relative arm movements for each pair does then allow a check to be made on the extensometer and specimen alignment. As with extensometer designs, a variety of strain-measuring techniques can be used. In general, dial gauges and mirror (Martens type) techniques are declining in popularity since the development of linear variable differential transformers (LVDT) or the more accurate differential capacitance transducers which can be linked to automatic data logging and analysis systems giving a digital and/or graphical record of the strain/time behaviour with any preset interval between individual readings.

During the last decade or so, computer-controlled servo-hydraulic and servo-mechanical testing machines have become available. Although these versatile machines cost considerably more than even high-precision constant-stress equipment and will not therefore be adopted for routine data acquisition, they permit complex stress/temperature patterns to be imposed for simulation of the non-steady stress and temperature conditions frequently encountered in power stations, chemical plant and aero-engines.

2.3. Factors Affecting Accuracy and Reproducibility

While a high degree of precision can be achieved with specialised equipment, a variety of standards have been introduced to define acceptable levels of accuracy and reproducibility of data for quality control and engineering design. In general, these standards have concentrated on specifying testpiece configurations, permissible temperature variations, and procedures for calibration of loads and extensometers (e.g. BS 1610 and BS 3846). With most specifications, the initial stress must be accurate to $\pm 0{\cdot}5\%$ which can be met by use of verified loading weights and by annual checking of the loading system by means of calibrated proving rings or load cells. Extensometers are then graded according to their room-temperature calibrations, e.g. in tests to BS 3846 the maximum permissible errors at a strain of 0·1 are $\pm 0{\cdot}0002$, $\pm 0{\cdot}0004$ and $\pm 0{\cdot}0008$ for Classes A, B and C, respectively. However, most concern is usually expressed about temperature variations since accurate temperature control as well as uniformity along the gauge length are required for acquisition of reliable long-term data. Various heating systems can be employed depending on the type of test involved but BS 3500 stipulates temperature control to ± 3 K at temperatures up to 873 K, ± 4 K for temperatures from 873 to 1073 K and ± 6 K for temperatures within the range 1073–1273 K.

Temperatures are normally measured using platinum/platinum–rhodium thermocouples since they can easily be made and calibrated to better than ±0·5 K. Moreover, thermocouples have a low thermal capacity, can be located easily and accurately, and are compatible with data logging systems. For high-precision testing, temperature variations can be reduced to better than ±1 K. However, the temperature variations quoted in most standards are not as imprecise as they first appear. In fact, finite element methods can be used to show that the creep rate of a sample having a temperature variation of 6 K along the gauge length is only 1·11 times faster than that expected if the temperature is constant.

While rigid limits are imposed on many test parameters, several others can be more difficult to quantify, such as the axiality of the loading system and the skill involved in setting up individual tests. For this reason it is essential to ensure that the extensometry and strain measuring systems are performing satisfactorily by measuring the elastic modulus at room temperature, and preferably at the creep temperature, before commencement of each test. This can be accomplished by loading and unloading, provided only that the maximum stress employed is less than the creep stress in order to avoid prestraining the material prior to the test.

The importance of the imprecisely defined or undefined factors in the national and international standards can be illustrated by the variation in the data reported for ostensibly the same material in tests carried out in 57 different laboratories in nine countries. In tests to around 30 000 h on two steels specially produced to ensure uniformity, the rupture lives varied from 15 000 to 36 000 h, which is equivalent to a scatter in stress of ±10 %.[2] Even greater scatter, which can be as much as ±20 % in stress (or approximately an order of magnitude in rupture life), is found in the international data being accumulated for individual grades of low-alloy steel used for electricity generating plant and petrochemical installations. Yet this scatter is not exclusively a problem of testing procedures. Figure 6 illustrates the creep curves recorded at 250 MN m^{-2} and 838 K for two batches of $2\frac{1}{4}$Cr–1Mo steel tested using high-precision constant-stress equipment (Fig. 5). Both steels were produced in the normalised and tempered conditions, and both were within the composition ranges specified for $2\frac{1}{4}$Cr–1Mo steel. Moreover, for each steel, reproducibility trials at the same stress and temperature showed that the rupture lives were within ±25 %. However, the material having a carbon content of 0·12 % produced in the fully bainitic condition was characterised by longer rupture lives but considerably lower rupture ductilities than the steel containing 0·08 %C, which had a microstructure of ferrite with only 20 vol % bainite due to the

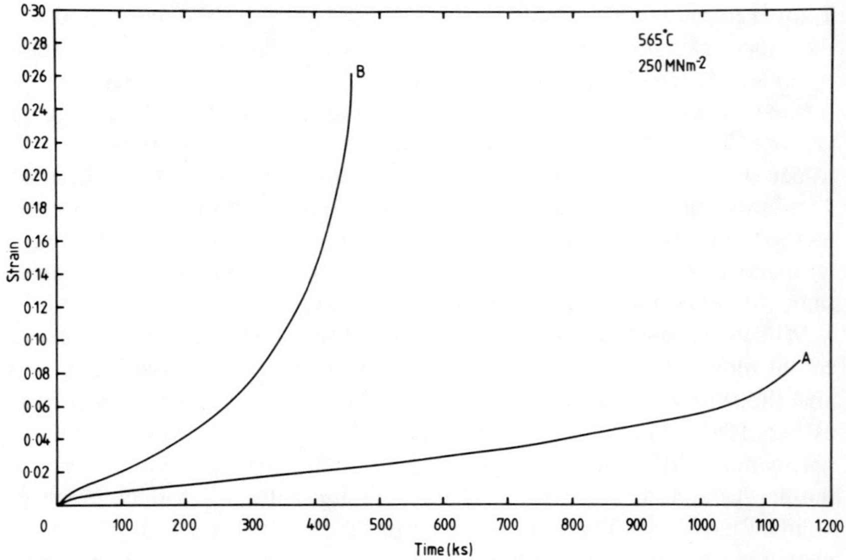

FIG. 6. Constant-stress creep curves recorded at 838 K and 250 MN m^{-2} for $2\frac{1}{4}$Cr–1Mo steel containing 0·12% carbon (A) and 0·08% carbon (B).

low cooling rates experienced. The large scatter bands associated with the creep and fracture data for steels which are within specification could then suggest that currently accepted compositional ranges should be even more restricted but, despite the control and flexibility offered by modern secondary steelmaking operations, this would pose major problems for steel producers.[3]

3. CREEP DATA ANALYSIS

When a stress, σ, is applied to a material maintained at constant temperature, T, the total true strain, ε, is made up of the virtually instantaneous strain on loading, ε_0, and the subsequent time-dependent or creep strain, ε_t. The basic information derived from a creep test is then the record of the variation of strain with time which, in turn, depends on the stress and temperature conditions imposed, i.e.

$$\varepsilon_t = f(t, \sigma, T) \tag{1}$$

Alternatively, the creep behaviour can be discussed in terms of the change in creep rate, $\dot{\varepsilon}$ ($= \mathrm{d}\varepsilon/\mathrm{d}t$), with either strain or time as

$$\dot{\varepsilon}_t = f(t, \sigma, T)$$

or

$$\dot{\varepsilon}_{\varepsilon} = f(\varepsilon, \sigma, T) \tag{2}$$

The development of valid equations relating stress–strain–time–temperature is sought for two principal reasons which, although they should be related, in practice are usually approached independently. The establishment of satisfactory constitutive laws which describe the macroscopic behaviour of metals and alloys is essential for the design of components and structures for service at elevated temperatures. In this case, provided only that the relationships offer an adequate basis for safe and economical design, it is not critical if the equations and the parameters which they contain are not understood in terms of the deformation and fracture processes taking place. Yet equations which correlate the patterns of behaviour for different metals and alloys are also a prerequisite for identification of the micromechanisms controlling deformation and fracture, and for clarification of the role of microstructural variables in determining creep and fracture resistance. In the final analysis these essentially macroscopic and microscopic approaches must be compatible but, as yet, a coherent solution has not emerged.

3.1. Polynomial and Exponential Descriptions of Creep Curves

The precise mechanism by which creep takes place, and hence the detailed strain/time response exhibited, depends both on the material being considered and on the ranges of stress and temperature being studied. Since the melting points of metals and alloys differ widely, it is therefore usual to discuss the creep behaviour displayed in relation to the fraction T/T_m, where T_m is the absolute melting point. Similarly, in order to facilitate comparisons between the strengths of different materials, the applied stress can be normalised by dividing by the appropriate elastic modulus.

The upper limit to the strength of crystalline solids is set by the theoretical shear strength which represents the stress required to cause shear of one plane of atoms over an adjacent plane in a perfect crystal. The theoretical shear strength is around $G/20$, where G is the shear modulus. Yet plastic deformation of metals and alloys almost invariably occurs on application of far lower stresses as a result of the presence of lattice defects such as vacancies and dislocations which can move either through the crystal lattice or along grain boundaries.

At temperatures sufficiently low that effects associated with diffusion can be ignored (i.e. $T < 0{\cdot}3T_m$), the application of tensile stresses less than the elastic limit results only in elastic strains, ε_e, of a magnitude given by Hooke's law (i.e. $\varepsilon_e = \sigma/E$, where E is Young's modulus). At higher stresses,

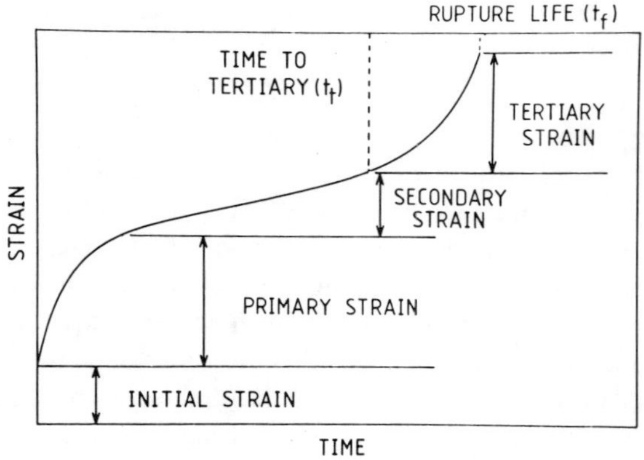

FIG. 7. Conventional representation of normal creep curves.

the resulting strains have both an elastic component and a plastic component with only the elastic strains being fully recoverable on removal of the stress. Under these conditions the creep strain varies logarithmically with time as

$$\varepsilon = \varepsilon_0 + \alpha_1 \log(\alpha_2 t + 1) \tag{3}$$

In this low temperature or logarithmic regime only limited creep strains are encountered (usually 1% or less) and the deformation processes taking place do not lead to eventual failure. For these reasons, logarithmic creep is of very limited engineering significance.

As the creep temperature increases towards $0{\cdot}3T_m$ and above, the shape of the creep curve departs increasingly from a logarithmic form and the creep rate decays more slowly than would be predicted from eqn (3). Indeed, during creep of most metals and alloys at temperatures above about $0{\cdot}4T_m$, after the initial strain on loading, the creep rate is reported to decrease gradually during the 'primary' stage until an apparently constant rate is attained during the 'secondary' stage. This secondary period does not continue indefinitely and, for most materials, the creep rate increases once again during the 'tertiary stage' which eventually leads to fracture. The principal features of this 'normal' creep curve and the terminologies usually adopted to describe this form of behaviour are illustrated in Fig. 7. However, it should be noted that this 'normal' curve shape is not always found and, more importantly, the relative importance of the different stages can vary for different materials and for different stress/temperature regimes.

The majority of attempts to describe this normal curve have

concentrated on the primary and secondary stages, with comparatively little attention directed to the tertiary curve despite its obvious importance in relation to fracture. In the classical work of Andrade[4] it was suggested that the time-dependent change in gauge length up to the 'steady-state' or 'secondary' condition can be described as

$$\ell = \ell_0(1 + \beta t^{1/3}) \exp(\kappa t) \tag{4}$$

with the β term describing the primary stage and the κ term ($=\dot{\varepsilon}_s$) quantifying the secondary rate. More usually, this type of equation is reduced to

$$\varepsilon = \varepsilon_0 + \beta t^{1/3} + \dot{\varepsilon}_s t \tag{5}$$

or, sometimes, to

$$\varepsilon = a_1 + a_2 t^{1/3} + a_3 t^{2/3} + a_4 t \tag{6}$$

Since polynomial descriptions of normal creep behaviour predict an infinite initial creep rate (i.e. $\varepsilon \to \infty$ as $t \to 0$) and do not lead to a 'steady' or constant creep rate during the secondary stage, a preference for exponential descriptions is often expressed as[5]

$$\varepsilon = \varepsilon_0 + \varepsilon_p[1 - \exp(-\gamma_p t)] + \dot{\varepsilon}_s t \tag{7}$$

where ε_p is the primary strain and γ_p relates to the rate of decay of the primary stage.

In order to describe the tertiary stage, polynomial equations such as[6]

$$\varepsilon = \varepsilon_0 + b_1 t^{1/3} + b_2 t + b_3 t^3 \tag{8}$$

allow the whole curve to be quantified by determinations of the magnitudes of the parameters b_1, b_2 and b_3 for any specified stress and temperature. Similarly, exponential expressions can be extended to include a tertiary term as[7]

$$\varepsilon = \varepsilon_0 + \varepsilon_p[1 - \exp(-\gamma_p t)] + \dot{\varepsilon}_s t + \varepsilon_T \exp[\gamma_T(t - t_T)] \tag{9}$$

where ε_T and γ_T define the shape of the tertiary period which begins after a time t_T.

All of these polynomial and exponential expressions envisage normal creep curves as being made up of three stages, i.e. primary, secondary and tertiary creep. An alternative description sees the shape of normal curves as a consequence of two competing events, decaying primary and accelerating tertiary processes, i.e. the secondary stage is then merely the period of inflection when the decaying rate during primary creep is offset by the acceleration due to the tertiary creep processes, giving an apparently 'constant' rate.[8] A polynomial expression compatible with this concept is[9]

$$\varepsilon = \varepsilon_0 + c_1 t^{\delta_1} + c_2 t^{\delta_2} \quad (10)$$

where $\delta_1 < 1$ and $\delta_2 > 1$. However, the problems inherent with polynomial equations can again be avoided by use of the exponential form[8] as

$$\varepsilon = \theta_1[1 - \exp(-\theta_2 t)] + \theta_3[\exp(\theta_4 t) - 1] \quad (11)$$

In this case, θ_1 and θ_3 act as scaling parameters which control the extent, with respect to strain, of the primary and tertiary stages of creep. θ_2 and θ_4 then determine the curvature of the primary and tertiary periods since increasing θ_2 and θ_4 rapidly increases the deceleration in creep rate during primary creep and the acceleration during tertiary creep.

3.2. Creep Data Fitting

For all of the polynomial and exponential expressions devised to quantify the shapes of normal creep curves, rough estimates of the parameters involved can be obtained graphically but good estimates require the use of suitable numerical parameter estimation procedures. By way of example, the full numerical procedures, the actual computer program listing, and the complete operating instructions necessary to determine the magnitudes of the θ terms in eqn (11) have been detailed elsewhere.[8] In this particular case, a least-squares method was adopted and the values of the θ terms were derived using between 100 and 200 strain/time readings recorded during high-precision constant-stress tests carried out for $\frac{1}{2}$Cr–$\frac{1}{2}$Mo–$\frac{1}{4}$V ferritic steel. The numerical solution involved a guarded Newton–Gauss procedure for minimising the least-squares function which was found to converge to unique values of the θ parameters. The least-squares method then has the added advantage that the confidence intervals for each parameter can be obtained in a straightforward way during the analysis.

Similar procedures can be adopted for all equations and it would seem reasonable to assume that a decision as to the most appropriate expression could be based on the accuracy of fit achieved for actual high-precision creep curves. However, within acceptable limits, the accuracy of fit as judged by the calculated standard errors is very similar for all equations. Selection of equations is therefore made by consideration of the physical significance which can be attached to the parameters in each expression. For example, computer analysis of high-precision curves using the polynomial in t (eqn (6)) shows that an excellent degree of fit can be achieved but that some coefficients can be negative and they rarely display any systematic variation with, say, stress or temperature. Consequently, while a perfectly acceptable description can be given for the strain/time behaviour in an individual test, this type of expression has limited value for interpolation or extrapolation of data, and for discussion of the deformation mechanisms governing creep behaviour. In general, these types of limitation are characteristic of polynomial expressions, hence the preference for exponential descriptions of creep curve shape.

In a similar manner, both eqn (9) and eqn (11) offer adequate representation of normal creep curves. A decision as to whether a definite 'steady-state' or 'secondary' stage exists, or whether the period of ostensibly constant rate is merely a period of inflection, cannot be resolved unambiguously on the basis of the 'degree of fit' achieved. For these reasons, the equation selected to represent curve shape is open to choice.

The problem of deciding upon the equation which best represents normal creep behaviour is usually avoided by the simple expedient of not analysing the curves at all. This solution has the added advantage of eliminating embarassment when the scatter in the strain/time readings is so large that meaningful parameter estimates cannot be derived. Instead, for reasons of choice or expediency, most theoretical studies of creep report only the secondary creep rates recorded. Irrespective of the precision of strain measurement, reasonable estimates of the secondary creep rate can be obtained relatively easily (at least in tests of fairly short duration). Moreover, measuring only the secondary creep rate appears to circumvent the difficulty encountered when the creep properties are defined in terms of the time to reach a specified strain of, say, 1 % or so. The relative importance of the primary, secondary and tertiary stages can change significantly with variations in stress and/or temperature. As a result, a strain of, say, 1 % may be reached early in the primary stage in high stress tests but only in the tertiary stage at low stresses. Although the time to attain a specified strain is essential for engineering design, most theoretical studies have therefore preferred to take the secondary creep rate as the parameter which characterises creep behaviour.

3.3. Representation of Standard Creep Parameters

With exponential descriptions of the shape of normal creep curves, and also with most of the polynomial expressions, direct relationships have been found between the secondary creep rate and the various terms used to quantify the primary and tertiary stages.[5,7] These relationships suggest that the mechanism by which deformation takes place is essentially the same throughout a normal creep curve. Most theoretical studies then attempt to identify the processes controlling creep behaviour by reference to the way in which the secondary or steady-state creep rate varies with stress and temperature.

At high temperatures (above $\sim 0{\cdot}4T_m$), the stress dependence of the secondary rate is generally described using a power law (often referred to as Norton's law) as

$$\dot{\varepsilon}_s \propto \sigma^n \tag{12}$$

where the stress exponent n represents the gradient of the $\log \sigma/\log \dot{\varepsilon}_s$ relationship. The actual magnitude of n can vary within very wide limits, ranging from unity to as much as 40 or more depending on the material and the test conditions. In the case of pure metals, n usually varies from about one at low stresses to about 4–6 at high stresses. Yet, even with pure metals, a rapid increase in n with increasing stress occurs at strain rates above $\sim 10^{-3}\,s^{-1}$. This phenomenon is usually described as the power-law breakdown and, at these higher stresses, the stress dependence of the creep rate is better expressed by an exponential form as

$$\dot{\varepsilon}_s \propto \exp(B\sigma) \tag{13}$$

where B is a constant.

Alternatively, it is possible to provide a more unified description which covers a very wide stress range, namely

$$\dot{\varepsilon}_s \propto [\sinh(B\sigma)]^n \tag{14}$$

which reduces to a power law at low stresses and an exponential at high stresses.[5]

The temperature dependence of the creep rate is normally well represented by an Arrhenius equation

$$\dot{\varepsilon}_s \propto \exp(-Q_c/RT) \tag{15}$$

where R is the gas constant when the activation energy for creep, Q_c, is expressed in units of $J\,mol^{-1}$ (R should be replaced by the Boltzmann constant, k, if Q_c is defined in units of $J\,atom^{-1}$). In the case of pure metals, Q_c is usually close to the activation energy for self-diffusion, although values approaching half those for self-diffusion can be found when diffusion occurs preferentially along dislocations or grain boundaries as the temperature decreases towards about $0{\cdot}4T_m$. Again, as with the n values reported using Norton's law, much higher values of Q_c (i.e. several times those for diffusion in the matrix) are frequently recorded for particle-strengthened alloys.[5,10]

Relationships similar to those used to describe the stress and temperature dependencies of the secondary creep rate are also widely adopted when considering stress-rupture behaviour since, for many materials over quite wide ranges of stress and temperature, the time to fracture is inversely proportional to the secondary creep rate.[11] At high stresses the deformation and fracture properties can then be presented as

$$\frac{1}{t_f} \propto \dot{\varepsilon}_s = A' \exp\left[-(Q_c - B\sigma)/RT\right] \tag{16}$$

and, at lower stresses, as

$$\frac{1}{t_f} \propto \dot{\varepsilon}_s = A\sigma^n \exp(-Q_c/RT) \tag{17}$$

The stress-rupture properties can therefore be described by exponential or power-law relationships but, as with the secondary creep behaviour, neither the $\sigma/\log t_f$ nor the $\log \sigma/\log t_f$ plots are linear over extended stress ranges. This non-linearity is also characteristic of various parametric relationships widely used to present engineering design data, as can be illustrated since both eqn (16) and eqn (17) can be rewritten as

$$\log t_f = \frac{C}{T} - D \tag{18}$$

When D includes the stress term (i.e. eqn (17)), this equation can be expressed as

$$D = \frac{C}{T} - \log t_f \tag{19}$$

with D being the parameter advocated by Orr *et al.*[12] Similarly, when C depends on stress (eqn (16)), the Larson–Miller relationship[13] is obtained as

$$C = T(D + \log t_f) \tag{20}$$

The aim of these parametric approaches is to derive a parameter, containing both the temperature and the rupture life, which can then be plotted as a function of stress. This type of approach can be useful for data collection and for interpolation. However, neither C nor D is linear over extended stress ranges as would be expected since neither the $\sigma/\log t_f$ nor the $\log \sigma/\log t_f$ plot is linear with increasing test duration. Consequently, this type of parametric relationship cannot be used reliably for extrapolation except over very limited time-spans.[14]

4. FACTORS AFFECTING CREEP AND FRACTURE BEHAVIOUR

When a tensile stress greater than the elastic limit is applied to a ductile metal at low temperatures, a finite plastic strain would be expected to occur by the generation and movement of dislocations. If the applied stress is below that needed to cause immediate fracture, the magnitude of this virtually instantaneous strain can be determined from the stress/strain curve for the material under the testing conditions imposed. When the tensile stress is maintained at these low temperatures, after the initial specimen extension on loading, the subsequent time-dependent strain accumulates in a 'logarithmic' manner (eqn (3)). The creep processes taking place under these conditions can be envisaged most easily by considering that, during the initial specimen extension, the material strain hardens by the generation and movement of dislocations until the flow stress becomes equal to the applied stress. Continued movement of any dislocation is then opposed by the long-range stress fields of the surrounding dislocations and also by short-range events such as the need for the glide dislocations to cut through forest dislocations intersecting the glide plane, the resistance to glide caused by jogs on the dislocations, etc. Creep then occurs as a result of thermal fluctuations helping the applied stress to continue dislocation movement. At low temperatures the thermal fluctuations enable the dislocations to overcome only short-range obstacles. Since the events requiring the least thermal energy will take place most rapidly, the creep rate will decrease continuously with time (eqn (3)). As the creep temperature increases towards $0{\cdot}4T_m$ and above, the creep curve departs increasingly from this logarithmic form because diffusion-related recovery processes allow rearrangement and annihilation of dislocations to occur continuously, i.e. the recovery processes gradually counterbalance the strain hardening associated with the generation and movement of dislocations until a 'steady-state' is achieved (Fig. 7).

A steady-state situation can also be attained when time-dependent deformation occurs by stress-directed vacancy flow at stresses too low for dislocation processes to be significant. A crystal in thermal equilibrium contains a concentration of vacancies, C_{eq}, given by

$$C_{eq} = \exp(-Q_f/RT) \tag{21}$$

where Q_f is the energy required to form a vacancy. External surfaces and grain boundaries are usually considered to be easy sources and sinks for vacancies. Each time a vacancy is produced at boundaries experiencing a tensile stress, the work done by the stress is of the order of $\Omega\sigma$, where Ω is the volume of a vacancy. The energy to form a vacancy at a boundary perpendicular to the tensile axis is therefore reduced from Q_f to $(Q_f - \Omega\sigma)$, so the local vacancy concentration is greater than the equilibrium value. Conversely, the vacancy concentration near a boundary under a compressive stress will be less than the equilibrium. Under the vacancy concentration gradient established in a pure crystalline metal of average grain diameter d, the material will extend in the direction of the applied tensile stress as the vacancies migrate (causing a counterflow of atoms) from boundaries perpendicular to the tensile axis to those parallel to the stress axis. The vacancy transfer can occur across the grains and around the grain boundaries so the overall creep rate is given by the sum of these vacancy fluxes as

$$\dot{\varepsilon} = C\frac{\sigma\Omega}{kT}\frac{D_L}{d^2}\left(1 + \frac{\pi\delta}{d}\frac{D_{gb}}{D_L}\right) \tag{22}$$

with $C \simeq 14$.[15] In this expression, δ is the grain-boundary width, and D_L and D_{gb} are the lattice and grain-boundary self-diffusion coefficients, respectively. In this way, when

$$\frac{\pi\delta}{d}\frac{D_{gb}}{D_L} < 1$$

the deformation rate is dependent upon diffusion through the lattice, a process referred to as Nabarro–Herring creep.[16,17] Similarly when

$$\frac{\pi\delta}{d}\frac{D_{gb}}{D_L} > 1$$

the creep rate is determined by grain-boundary diffusion, which is usually termed Coble creep.[18]

A variety of dislocation and diffusional creep theories have therefore been developed to explain creep behaviour. Indeed, for any given material, variations in the creep properties exhibited under different test conditions are usually interpreted on the basis that different mechanisms become dominant in different stress/temperature regimes.[19]

4.1. Deformation Mechanism Maps

The relative importance of dislocation and diffusional processes during creep at high temperatures is usually inferred from measurements of the dependence of the secondary creep rate on stress, temperature and grain size, using an expression of the general form

$$\dot{\varepsilon}_s = A(1/d)^m\sigma^n \exp(-Q_c/RT) \tag{23}$$

Most creep processes cannot be quantified sufficiently well to allow the relevant deformation mode to be ascertained from a comparison of measured and predicted creep rates. As a result, the dominant creep process is generally identified by comparing measured values of the grain size exponent (m), the stress exponent (n) and the activation energy for creep (Q_c) with the theoretical values expected for specific creep mechanisms.[20]

Dislocation and diffusional creep processes contribute independently to the overall creep rate, i.e. the creep rate is determined as the sum of the rate due to dislocation processes and that due to diffusional creep. When two independent mechanisms are associated with stress exponents of n_1 and n_2 (with $n_1 < n_2$), the process having the larger exponent is dominant at higher stress levels. While mechanisms based on the generation and movement of dislocations are normally characterised by n values of 3 or more, diffusional creep theories predict stress exponents close to unity (eqn (22)). For this reason, changes in the gradient of the $\log \sigma / \log \dot{\varepsilon}_s$ relationships from $n > 3$ at high stress levels to $n \simeq 1$ at low stress levels have usually been taken as evidence of a change in rate-controlling mechanism from dislocation to diffusional processes.

In the $n \simeq 1$ regime, measurements of Q_c (normally derived from the gradients of $\log \dot{\varepsilon}_s/(1/T)$ plots) are frequently used to determine whether diffusional flow occurs predominantly through the lattice or along grain boundaries. With pure metals, Q_c values close to those for lattice diffusion should be indicative of Nabarro–Herring creep, whereas Q_c values approaching half those for lattice diffusion would be associated with Coble creep (eqn (22)). Since the process having the larger activation energy would be favoured at higher temperatures, a transition from Nabarro–Herring to Coble creep could be expected with decreasing temperature in tests undertaken at low stresses. However, a similar decrease in Q_c value with decreasing temperature would also occur when creep occurs by dislocation processes. In this case, the lower activation energy would be due to preferential diffusion along dislocations (usually termed 'pipe diffusion'). In both the dislocation and diffusional regimes, Q_c measurements account for differences in the creep characteristics observed at high temperatures when the creep rate is determined by lattice diffusion and at intermediate temperatures (towards $\sim 0{\cdot}4T_m$) when preferential diffusion paths are favoured.

Supplementary evidence as to the rate-controlling process has also been obtained from measurements of the dependence of $\dot{\varepsilon}_s$ on the grain size of the material at constant stress and temperature. In general, dislocation processes are considered to be independent of grain size whereas a marked

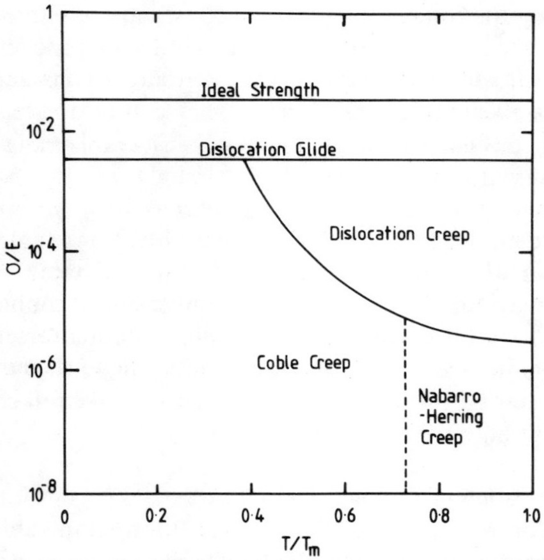

FIG. 8. Schematic deformation mechanism map for a polycrystalline metal.

grain-size dependence would be anticipated when diffusional creep processes are dominant (eqn (23)), i.e. m values of 0, 2 and 3 would be expected for dislocation creep, Nabarro–Herring creep and Coble creep, respectively.[20]

By these procedures, measurements of n, m and Q_c appear to offer a satisfactory means by which the rate-controlling process can be identified. However, the dominant creep mechanism and hence the creep properties would be expected to differ in different stress/temperature regimes. For this reason, considerable emphasis has recently been placed on the presentation of a compendium of information as 'creep diagrams'[21] or, in more detailed style, as 'deformation mechanism maps'.[19,22] These maps present the deformation characteristics of a material in the form of a stress/temperature diagram, with axes of σ/E and T/T_m (Fig. 8). For any specified creep conditions, the contribution of each creep mechanism to the overall creep rate is calculated from the appropriate constitutive equation relating stress, temperature and strain rate for each process (e.g. eqns (22) and (23)). Maps may then be constructed using actual data (i.e. using measured values of n, m and Q_c in eqn (23)) or by means of the relevant theoretical equations for each process (e.g. eqn (22)). The ranges of stress and temperature over which each process predominates can therefore be identified, with the boundaries between adjacent fields (heavy lines in

Fig. 8) indicating the conditions under which two mechanisms contribute equally to the creep rate.[19] Furthermore, at a given stress and temperature, a material exhibits a fixed creep rate so that contours of constant creep rate (or of time to a given creep strain) can be included on the map.

Deformation mechanism maps allow a considerable range of information to be presented in a compact, highly visual form. It must, however, be emphasised that the accuracy of a map is only as good as the experimental data and/or the theoretical equations employed for construction. Also, maps represent only 'steady-state' creep. The logarithmic behaviour observed at low temperatures cannot even be included since a 'steady' creep rate is not attained. Moreover, it should be recognised that the 'secondary' or 'steady-state' stage, if present at all, usually accounts for a relatively small proportion of the full creep 'curve' recorded even at high temperatures.

4.2. The Interrelation of Creep and Fracture

Many materials which are relatively ductile at low temperatures can fail in a brittle intergranular manner under high temperature creep conditions. Indeed, when crystalline solids are deformed in tension at a specified strain rate, a plot of total strain to fracture (or reduction in area at fracture) as a function of temperature frequently shows a 'ductility minimum'. This phenomenon is simply a manifestation of the fact that the fracture mechanism depends critically on the material considered and on the test conditions imposed.[23,24] For example, at low temperatures failure under tensile stresses may occur by cleavage, by ductile fracture, or by necking or shear. At temperatures relevant to operation of high temperature components and structures, fracture can take place in a ductile transgranular or intergranular manner, or as a result of the formation of intergranular wedge cracks or cavitation (Fig. 9). At high temperatures and rapid strain rates, dynamic recrystallisation can allow materials to deform

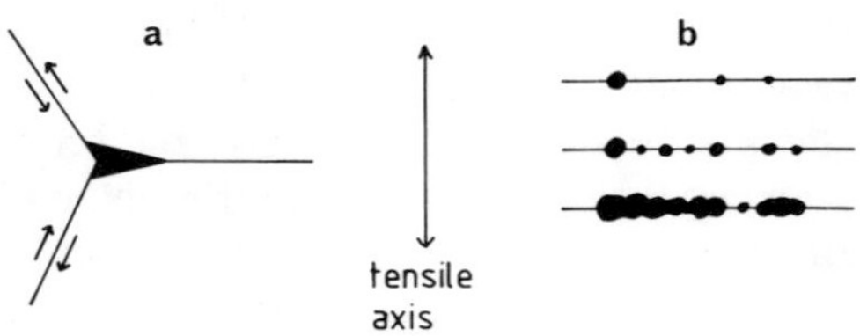

FIG. 9. Schematic illustration of: (a) the formation of triple-point or wedge cracks, and (b) the growth and link-up of cavities on grain boundaries normal to the tensile axis.

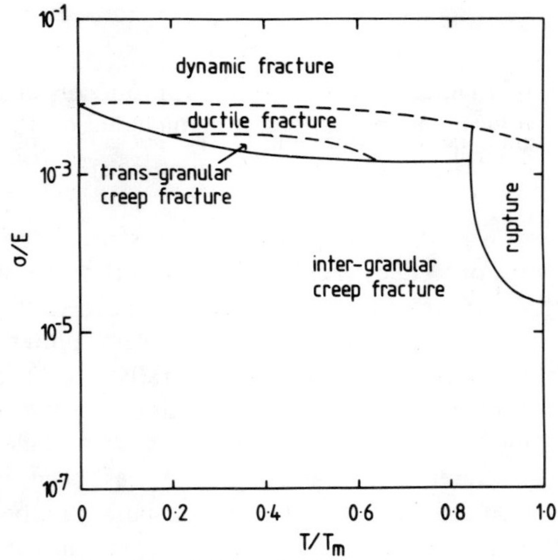

FIG. 10. Schematic fracture map.

extensively with failure eventually due to the specimen necking down to virtually zero cross-section, usually termed rupture.[23]

For reasons analogous to those prompting the construction of deformation mechanism maps (Fig. 8), attention has been directed[23] to the development of 'fracture maps' which display the relevant failure modes for a specific material in terms of a stress/temperature diagram (Fig. 10). Under any prescribed test conditions, the mechanism resulting in fracture in the shortest time or after the lowest strain determines the fracture characteristics of the material. The 'ductility minimum' exhibited by many materials then coincides with conditions giving rise to intergranular creep fracture, i.e. higher ductilities would be recorded when fracture occurs in a relatively ductile manner at low temperatures and also when dynamic recrystallisation results in rupture at high temperatures (Fig. 10).

Intergranular creep fracture is generally associated with the formation of wedge cracks or with the nucleation, growth and eventual coalescence of grain-boundary cavities (Fig. 9). Wedge cracks form as a result of the stress concentrations generated at triple grain junctions as grain-boundary sliding takes place during creep. The stress required to create a wedge crack, σ_w, depends both on the surface energy of the material, γ, and on the average grain size, d, as[25]

$$\sigma_w = \sqrt{\left[\frac{6\gamma G}{\pi(1-\nu)d}\right]}$$

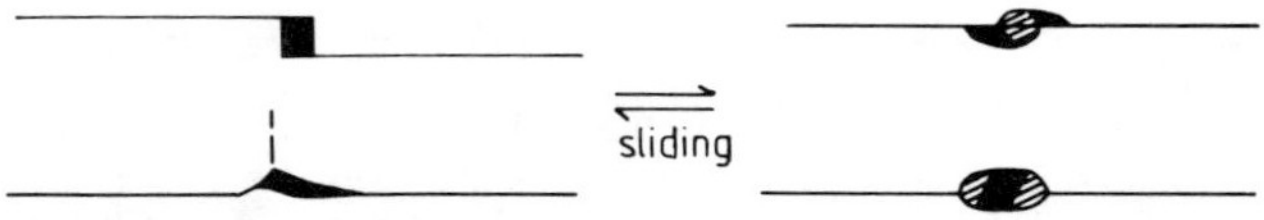

FIG. 11. Diagrammatic representation of the nucleation of grain-boundary cavities at ledges, cusps and particles.

where v is Poisson's ratio. Unless the applied stress is greater than this value, triple-point or wedge cracking will not initiate. Consequently, this form of intergranular cracking is observed only at relatively high stresses. During long-term creep exposure intergranular failures are normally associated with cavitation. It is then generally accepted that cavities nucleate as a result of the stress concentrations developed when grain-boundary sliding is impeded at ledges, cusps or particles (Fig. 11). In contrast, the mechanism of cavity growth has been the subject of controversy for several decades. The broad-based support for models involving diffusional growth of cavities by direct vacancy absorption from the surrounding grain boundaries experiencing a tensile stress[26,27] has now faded with the acceptance of the evidence supporting strain-controlled cavity development. For example, density measurements have shown that cavities can be nucleated early in the creep life and that the total cavity volume increases linearly with increasing creep strain.[28] More importantly, cavity growth by direct vacancy absorption would inevitably lead to a temperature dependence of the rupture life related to the activation energy for grain-boundary diffusion.[26,27] Yet for many materials over quite wide ranges of stress and temperature the rupture life is inversely proportional to the creep rate.[11] Early results showing a low activation energy for fracture[26] were then taken as evidence for cavity growth controlled directly by grain-boundary diffusion, not realising that low Q_c values (and hence low activation energies for fracture) were relevant only at temperatures around $\sim 0{\cdot}4T_m$. However, despite the current acceptance of strain-controlled cavitation, no general agreement has so far been reached on the detailed mechanism of cavity growth. As a result, different investigators have favoured models ranging from those dependent on growth associated with localised deformation in the grain-boundary regions[28,29] to those assuming growth by vacancy absorption with the growth rate constrained by the rate at which deformation can occur in the surrounding material.[30,31]

4.3. Creep Damage Accumulation

With many pure metals and relatively simple alloys, intergranular cavities and cracks develop throughout the creep life. In these cases the accumulation of damage can be readily quantified. Indeed, by comparing the extent of the intergranular damage in a component withdrawn from service with the known damage accumulation for the relevant material, an estimate of the remnant life of the component may be obtained.[32] However, this approach is not universally applicable. For example, with many commercial creep-resistant alloys microstructural examination of samples cooled under load after various creep strains shows that significant levels of intergranular damage are often detectable only very late in the creep life.[8] A comprehensive understanding of creep fracture must obviously cater for all types of damage accumulation so that the behaviour of different categories of material can be explained. In attempting to achieve this rationalisation, it seems reasonable to start by accepting that intergranular damage accumulation is strain-controlled and then assessing the various forms of damage in relation to the overall creep curve for the appropriate material.

In view of the form of the normal creep 'curve' (Fig. 7) and the well-established change in the relative importance of the various stages of creep as a function of test conditions,[33] it must be considered surprising that the vast majority of academic studies have concentrated on 'steady-state' creep behaviour and on the detailed processes responsible for intergranular fracture. While relatively little attention has been devoted to primary creep, even less emphasis has been directed towards the tertiary stage. Yet, under high temperature creep conditions, the first indication of eventual fracture is usually the acceleration in creep rate observed when the tertiary stage begins. Furthermore, at the low stress levels encountered in service, the creep curve exhibited by most commercial creep-resistant alloys can be considered as consisting of only a tertiary stage, i.e. an acceleration in creep rate is apparent almost from the commencement of a long-duration creep test.[8,33]

The tertiary stage is often dismissed as a consequence of the gradual increase in stress which occurs as the cross-sectional area of the testpiece decreases with the constant-load test procedures widely adopted. Although the stress increase inevitable with constant-load equipment will obviously result in an escalation of the creep rate recorded for ductile materials (Fig. 2), a well-defined tertiary stage is almost invariably found under constant-stress conditions (Fig. 6). At constant stress, the acceleration in creep rate may be attributable to:

(a) The development of grain-boundary cavities and cracks to a size sufficient to influence the deformation rate.

(b) Necking of the testpiece.
(c) Microstructural instability, such as grain growth or recrystallisation in single-phase materials or increases in the size and spacing of precipitates as ageing continues during creep exposure of precipitation-hardened alloys.

With pure metals and relatively simple alloys tested under conditions ensuring microstructural stability, when the total volume of cavities and cracks increases linearly with strain throughout most of the creep life,[28] models can be developed which account for the precise form of tertiary curve displayed.[34] However, with materials such as the $\frac{1}{2}$Cr–$\frac{1}{2}$Mo–$\frac{1}{4}$V ferritic steel widely used in UK electricity generating plant relatively few cavities and cracks are discernible until necking takes place late in the creep life.[8] Examination of fractured testpieces then reveals that the incidence of cracking decreases rapidly with distance from the actual fracture surface, with little evidence of intergranular damage outside the necked portion of the fractured specimen. These observations indicate that, for materials such as $\frac{1}{2}$Cr–$\frac{1}{2}$Mo–$\frac{1}{4}$V steel, intergranular cavities and cracks nucleate only after relatively large creep strains have occurred or as a result of the triaxial stresses generated when necking takes place after the required creep strains have been accumulated. With the $\frac{1}{2}$Cr–$\frac{1}{2}$Mo–$\frac{1}{4}$V steel tested at high stresses, the onset of the tertiary stage appears to coincide with the stage at which necking becomes apparent. However, as the test duration increases with this type of alloy, the point at which an increase in creep rate is detectable occurs at a progressively earlier fraction of the total creep life, i.e. the tertiary stage begins well before necking and the associated crack development is apparent. Under these conditions, the initial acceleration in creep rate is attributable to the gradual loss of creep strength as the size and spacing of the carbide dispersion takes place with increasing creep exposure. Even so, fracture eventually occurs as a result of necking and crack formation when the necessary creep strains have been accumulated.[8]

The relative importance of the various causes of tertiary creep, and hence the precise form of the tertiary curve, obviously depends on the material and on the test conditions. It is then interesting to note that the actual rupture life is determined largely by the shape of the tertiary curve, i.e. since the creep rate is accelerating rapidly with increasing time (or strain), for a given tertiary curve, even a relatively large change in creep ductility will cause only a comparatively small variation in rupture life (Fig. 6). On this basis, the fracture characteristics of a material under any prescribed set of test conditions would appear to be characterised more usefully by the development of procedures able to quantify the creep curve shape as a function of stress and temperature, which would then allow the rupture life to be derived from a knowledge of the factors governing creep ductility. It is, in fact, this approach which is now being pursued in order to provide a solution to one of the major problems encountered in the design of components and structures for high temperature service, namely the accurate prediction of the long-term creep and fracture properties of engineering materials by extrapolation of short-term test data.[8]

5. MATERIALS FOR HIGH TEMPERATURE SERVICE

The service lives acceptable for components and structures operating at elevated temperatures vary considerably depending on the particular application envisaged. Components such as aero-engine turbine blades are usually expected to operate satisfactorily for periods of between 10 000 and 20 000 h. In contrast, the design lives of the pipework and other major components for electricity generating plant are normally 100 000 h, although considerable economic benefits may be derived if 250 000-h lives (almost 30 years) can be achieved safely. Materials which are capable of withstanding the anticipated service conditions for the relevant times, and which can be fabricated to the shapes required at a suitable cost, are then essential for efficient operation and for avoidance of unscheduled and often expensive replacement or repair of components within the stipulated design lives. Yet, despite the impressive advances which have been made in alloy development and manufacture, the maximum operating temperatures acceptable in service must still frequently be limited to match the performance of the available materials.

5.1. Creep-Resistant Materials

With even an imprecise knowledge of creep mechanisms, the factors likely to result in improved creep strength can be identified qualitatively. For example, certain elements in solid solution may increase creep resistance by virtue of their low diffusivities or by lowering the stacking fault energy of the base material which retards dislocation glide and recovery processes. Moreover, when two or more alloying elements are present in solid solution, their combined effects can be much greater than the sum of their individual contributions. In this context, there is now ample evidence to show that the creep strength of simple mild or carbon/manganese steel is due largely to an interactive solid-solution hardening effect arising from the simultaneous presence of dissolved manganese and nitrogen and, to a lesser extent, manganese and carbon.[35] With this group of commercial steels, which incidentally exhibit creep strengths considerably in excess of that for pure iron, a relatively unimportant contribution to strength is made by the other elements present in solid solution, e.g. copper, phosphorus, silicon, etc. The importance of this type of interactive solid-solution strengthening, compared with even the development of a fine iron–manganese nitride dispersion, can be illustrated by results obtained for an Fe–1·6 % Mn–0·045 %N alloy.[36] This particular alloy has manganese and nitrogen contents sufficient to allow nitride formation by ageing or creep testing at ~750 K, yet the creep resistance decreased rather than increased during the period when precipitation occurred.[36]

Although the case of manganese-containing steels does illustrate the contribution to creep resistance which can be obtained by solid solution strengthening, most commercial creep-resistant alloys derive their high temperature strength from the presence of a fine dispersion of second-phase particles. Indeed, a sufficient volume fraction of closely spaced particles can inhibit creep, irrespective of whether deformation occurs by dislocation or diffusional creep processes. Since the strengths of particle-hardened alloys depend on maintaining a fine dispersion, it is thus essential to minimise particle coarsening at elevated temperatures. To ensure a low interparticle spacing, certain steels rely on a succession of phases (usually carbides) forming sequentially, so that each new precipitate provides the resistance to deformation as the earlier precipitate coarsens.[37] Alternatively, since the driving force for particle growth is provided by the precipitate/matrix interface energy, microstructural stability can be enhanced by control of the alloy composition to minimise the interfacial energy. Even greater resistance to growth is found for alloys containing a dispersion of highly insoluble particles. Examples of this type of alloy are sintered aluminium powder which has a dispersion of alumina particles in an aluminium matrix (SAP) and the thoria- or yttria-dispersed nickel or nickel–20 % chromium alloys (referred to as TD-nickel, TY-nickel, TD-nichrome, etc). In these cases the particle stability is derived not from a low interfacial energy but from the extremely low solubility of the dispersed phase. Growth of a large particle at the expense of surrounding smaller particles is then restricted by the very slow transport rates of the appropriate elements through the matrix.

These considerations illustrate that a qualitative explanation can be provided for the way in which various microstructural features affect creep

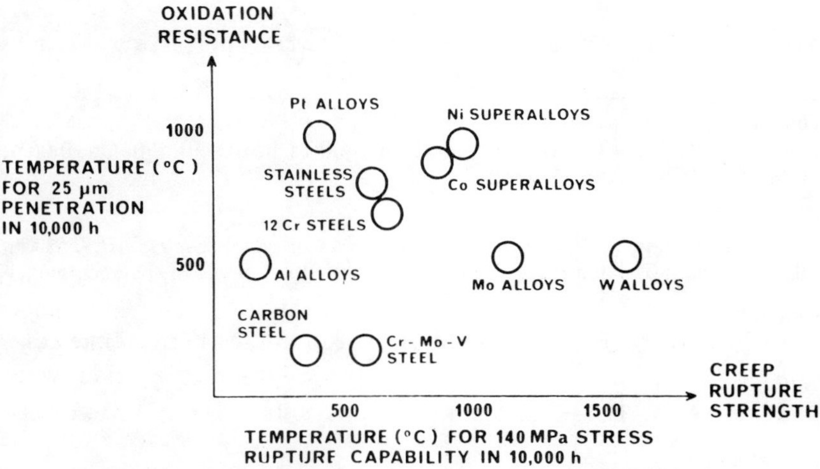

FIG. 12. Creep-rupture strength and oxidation resistance of various alloy systems.[38,39]

resistance, but the exact compositions required in order to maximise the creep strength of any base material cannot be predicted theoretically. Alloy development has therefore proceeded in an empirical, but highly systematic, manner. Even so, dramatic improvements have been made in the performance of a wide range of alloys over recent decades. In any case, creep resistance alone is rarely sufficient. If it were then the variety of ceramic materials already available would meet all existing requirements. Unfortunately, assuming that components could be fabricated from the appropriate engineering ceramics, the brittleness of this class of materials poses major design problems. Even with metallic materials it is necessary to ensure that the requisite creep strength is combined with adequate ductility so that, for example, the alloy can deform rather than fracture in regions of high stress concentration; this allows the stress to be redistributed onto the remainder of the component cross-section.

Satisfactory mechanical properties must also be combined with adequate resistance to environmental degradation in order that the material can withstand the hostile conditions encountered in many high temperature applications. Currently, nickel-base and cobalt-base alloys offer the best combination of strength and oxidation resistance at high temperatures (Fig. 12), with the nickel-base alloys being most widely adopted because of the price volatility of cobalt.[38,39] These alloys are capable of sustaining considerable loads for many thousands of hours at temperatures

TABLE 1

TYPICAL COMPOSITIONS OF SELECTED NICKEL-BASE SUPERALLOYS (ALL COMPOSITIONS WEIGHT PERCENT)

Nimonic 80A	19·5 Cr, 1·4 Al, 2·4 Ti, <0·05 Co, 0·02 Mn, <0·15 Si, 0·003B, 0·02 Cu, 0·06 C, 0·005 S, 0·19 Fe, balance nickel
Nimonic 105	14·85 Cr, 4·79 Al, 1·23 Ti, 20 Co, 4·98 Mo, 0·04 Mn, <0·15 Si, 0·006 B, 0·01 Cu, 0·125 C, 0·004 S, 0·095 Zn, <0·001 Ag, <0·001 Pb, <0·001 Bi, balance nickel
Nimonic 115	14·5 Cr, 5·0 Al, 3·8 Ti, 13·3 Co, 3·3 Mo, 0·016 B, 0·15 C, 0·045 Zr, balance nickel
IN100	10 Cr, 5·5 Al, 4·7 Ti, 15 Co, 3 Mo, 1·0 V, 0·014 B, 0·18 C, 0·06 Zr, balance nickel
Mar M002	9 Cr, 5·5 Al, 1·5 Ti, 10 Co, 10 W, 0·5 Mo, 0·2 Mn, 0·2 Si, 0·015 B, 0·05 Zr, 0·5 Fe, 2·5 Ta, 1·5 Hf, 0·1 Cu, 0·15 C, balance nickel

approaching 1300 K. Yet the creep and environment resistances of even the best superalloys do not approach those of modern engineering ceramics such as the silicon nitride and carbide type materials now being developed. Although the use of this type of ceramic will necessitate the evolution of new design procedures, the potential of these materials is such as to ensure their consideration for a range of high-performance applications.[40]

5.2. Nickel-Base Superalloys

Modern nickel-base superalloys have evolved from the single-phase nickel–20% chromium alloy, relying on the formation of an adherent Cr_2O_3 scale to confer resistance to high temperature degradation.[41,42] The developments which have taken place to reach the current range of multi-element, multi-phase alloys can be illustrated by reference to the Nimonic®* series of alloys. It was found that the creep strength of the basic nickel–chromium alloy could be increased considerably by the addition of aluminium and titanium to give Nimonic 80, which was later modified to Nimonic 80A (Table 1). The high temperature capability was later improved by the addition of ~20% cobalt to give Nimonic 90 and then by increasing the levels of aluminium and titanium, and introducing molybdenum, to give Nimonic 105 and 115 (Table 1).

* Nimonic® is a registered trade mark of Henry Wiggin and Co. Ltd, Hereford, UK.

The addition of aluminium and titanium permits the development of ordered $Ni_3(Al, Ti)$ or γ'-type precipitates which are isomorphous with the fcc γ matrix. Indeed, the lattice parameter of the γ'-phase may differ by less than 1% from that of the matrix. The resulting low interfacial energy extends the temperature capability of this form of alloy by about 200 K beyond that expected for an equivalent volume fraction of precipitate having a high interfacial energy. Niobium additions can also enter the γ'-phase although, if the niobium content exceeds ~4%, a separate Ni_3Nb phase can precipitate. Part of the titanium and niobium can also exist as carbides, as does a proportion of the molybdenum, tungsten and tantalum present in the higher strength alloys (Table 1). Although the formation of carbide particles contributes to the creep strength, the molybdenum, tungsten and tantalum serve primarily to promote solid-solution strengthening at high temperatures. Cobalt also results in a slight solid-solution hardening but is added mainly because it lowers the solubility of aluminium and titanium in the matrix. Boron, zirconium and, more recently, hafnium may also be added because of their contribution to strength and ductility. The exact roles of these and other elements present in nickel-base superalloys, the factors affecting phase stability, and the detailed heat treatments necessary to optimise microstructure and properties have been detailed elsewhere.[41,42] However, the improvements in creep strength which have been achieved by control of the composition and fabrication procedures are apparent from Fig. 13, which shows the high temperature capability of the various alloys listed in Table 1.

The compositions listed in Table 1 provide examples of both wrought and cast superalloys. Essentially, the increase in strength achieved for the wrought alloys was obtained by increasing the aluminium and titanium contents. These elements have the additional effect of decreasing the liquidus and solidus temperatures while simultaneously raising the solution temperature of the γ'-phase which restricts the permissible hot-working range of the higher strength alloys. For this reason, casting procedures were developed which, by obviating the need for forging operations, allowed even greater levels of alloying additions to be used, e.g. the alloys IN100 and Mar M002 in Table 1. The resulting microstructural complexity of these cast alloys can be inferred from Fig. 14. The superior creep strengths of these advanced cast alloys are then due largely to the high volume fractions of γ'-phase present, to increased solid-solution strengthening and to the large grain size obtained.[42] Further improvements in creep life and ductility have then resulted because of process innovations[43] which produce a columnar grain structure by unidirectional solidification (UDS

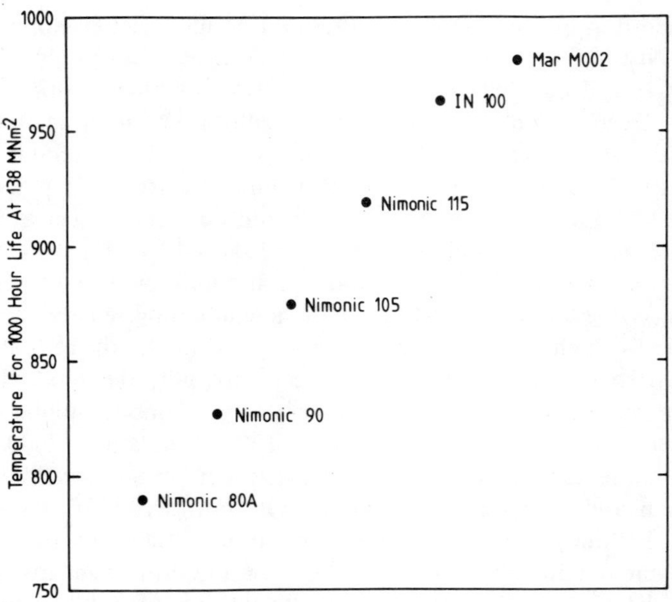

FIG. 13. Temperature for 1000 h creep lives at 138 MN m^{-2} for various nickel-base superalloys.

alloys) and even by manufacture of components in the form of single crystals (Fig. 15).

Superalloys strengthened by a γ'-phase cannot be used for prolonged periods at temperatures significantly above ~ 1320 K, even when coatings are employed for added oxidation resistance. For this reason, efforts are being directed towards the development of directionally solidified eutectic compositions which rely for their strength on alignment of carbide or intermetallic phases. Attention is also being focussed on 'mechanical alloying'. This procedure uses high energy milling techniques to incorporate non-metallic particles such as yttria into alloy powders. The resulting dispersion-strengthened powders can then be fabricated to shape and heat-treated to produce alloys having a dispersion of both γ' precipitates and insoluble particles. However, while developments offering only 10 K improvements in high temperature performance confer major benefits in critical applications such as aero-engine turbine blades, the fact that superalloys are now operating at around $0{\cdot}8T_m$ suggests that little scope exists for further dramatic improvements in superalloy capability.

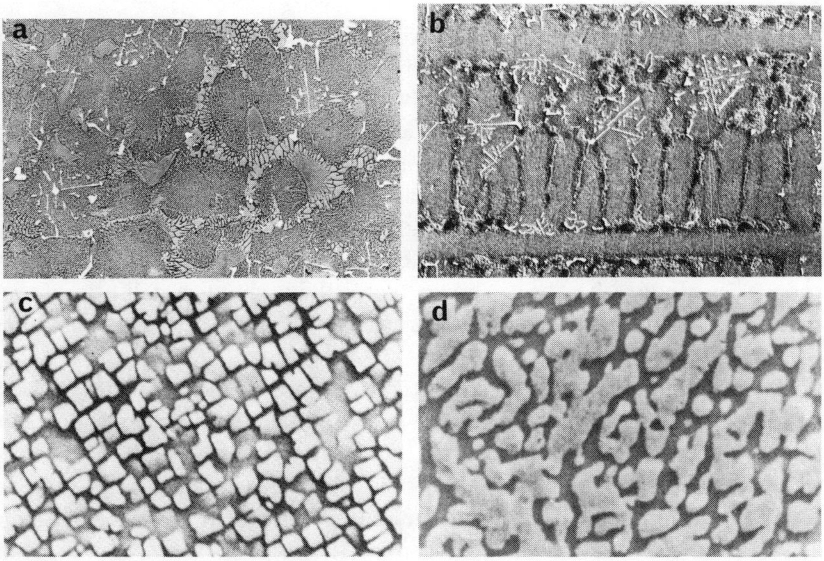

FIG. 14. Typical microstructures of Mar M002: (a) in the conventionally cast form and (b) in the directionally solidified condition. Changes in the γ' dispersion during creep are illustrated for cast superalloy IN 100, in Figs (c) and (d) which show the primary γ' precipitation before creep exposure and late in the tertiary stage of creep tests undertaken at 1223 K[45] ((a) and (b) $\times$ 100; (c) and (d) $\times$ 4000).

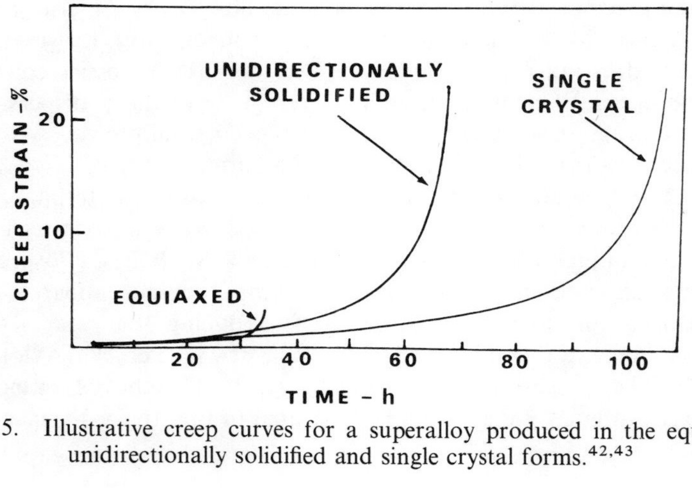

FIG. 15. Illustrative creep curves for a superalloy produced in the equiaxed, unidirectionally solidified and single crystal forms.[42,43]

5.3. Rejuvenation

In view of the cost and complexity of many components serving at elevated temperatures, methods which permit extension of their useful lives may offer considerable economic advantages. For this reason, numerous studies have been undertaken in an attempt to improve the creep lives of nickel-base superalloys by employing repetitive creep/reheat treatment cycles, i.e. a testpiece is crept for a time t_a, usually to the late secondary or early tertiary stage, reheat-treated and the test continued for a further period, t_a, and so on. While indefinite creep lives cannot be obtained, this type of procedure has been shown to result in significant life improvements, although it is necessary to devise specific heat-treatment schedules for individual superalloys.[44,45]

With Nimonic 80A the creep life can be improved by a factor of ~4 simply by using heat treatments which involve holding at the creep temperature under zero stress. In contrast, with the higher strength wrought alloys such as Nimonic 105 and 115 and with cast alloys such as IN100 life improvements were obtained only with heat treatments which dissolve the γ'-phase and then reproduce the original γ' dispersion. This behaviour can be interpreted directly in terms of the patterns of creep damage accumulation identified for different superalloys. In the case of Nimonic 80A grain-boundary cavities and cracks form relatively early in the creep life and grow throughout the creep test until crack link-up causes fracture. Since cavitation is responsible both for the onset of tertiary creep and for eventual fracture with this alloy, significant life improvements are obtained by periodically sintering-out the developing cavities before extensive cracking develops. With Nimonic 80A cavity sintering can be accomplished by annealing at the creep temperature. However, this procedure does not result in indefinite lives because successive creep and annealing operations gradually allow overageing of the γ' dispersion so that, as a progressive loss of creep strength occurs, failure will eventually take place in the interval between sintering periods.

Unlike the situation relevant to Nimonic 80A, the tertiary stage commences as a result of microstructural instability in long-term tests carried out for high strength superalloys (Fig. 14). With these alloys, cracks develop only late in the life when necking becomes apparent. Life enhancement must then be achieved by reducing the rate of strain accumulation and thereby deferring the time at which crack development initiates. The increase in life obtained by periodic reheat treatment to restore the original γ' dispersion is illustrated in Fig. 16 for the cast alloy IN100.

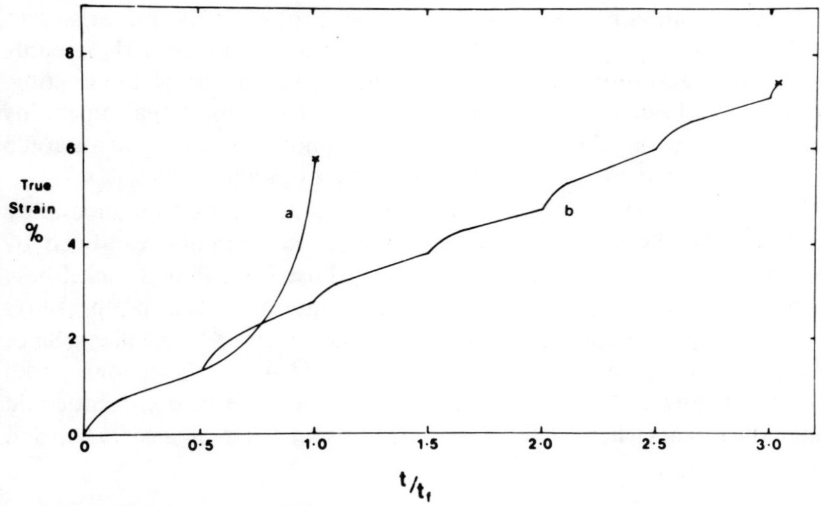

FIG. 16. Illustration of the improvement in creep life attainable for IN 100 by periodic reheat treatment.[45] Curve a shows the uninterrupted creep curve at 185 MN m^{-2} and 1223 K. Curve b indicates the life enhancement which can be achieved by periodic reheat treatment involving annealing at 1493 K for 3·6 ks followed by cooling at 0·1 K s^{-1}.

These considerations suggest that incorporating a hot isostatic pressing (HIP) operation into the rejuvenation schedule is unlikely to result in life improvements which are dramatically greater than those which can be achieved for high-strength superalloys by direct restorative heat treatments. Certainly, HIP of components such as turbine blades before service can maximise performance by eliminating the internal pores which can be present in cast superalloys. However, since internal cavities and cracks are not formed until late in the creep life of conventionally cast materials, the inclusion of HIP operations into the rejuvenative heat-treatment cycle would seem only a precaution. In any case, HIP processing will not eliminate surface cracks and, with the complex stress conditions encountered in turbine blades, failure can often originate at surfaces.

In the case of aero-engine components, a more serious limitation may be imposed by virtue of the fact that rejuvenative procedures applied late in the creep life appear to be ineffective.[44,45] Yet aero-engines are generally overhauled after fixed periods in service. In order to obtain a worthwhile life improvement, the overhaul interval would have to coincide with the period at which rejuvenation would be effective, i.e. the service interval would have to be less than half the expected life of the component under the operating conditions experienced. A further problem may then arise with coated hollow turbine blades. Satisfactory rejuvenation would probably have to be accompanied by stripping and replacement of the coating. Removal of the coating may also remove $\sim 25\ \mu m$ of the actual superalloy and, with thin-walled blades, the number of rejuvenative treatments could well be restricted by the associated reduction in section thickness.

While rejuvenation studies have tended to concentrate on superalloy applications, the patterns of creep damage accumulation exhibited by various steels have obvious parallels with those for different nickel-base alloys. Consequently, the principles of rejuvenation relevant to superalloys should be equally applicable to steels and other creep-resistant alloys. Since the complications associated with coatings and thin-wall sections would not be encountered, life enhancement by periodic reheat treatment could offer advantages when applied to components for electricity generation and chemical plant.

6. CONCLUSIONS

The procedures for creep and stress-rupture testing are well established and, if the relevant national and international standards are adhered to, few problems should be encountered in the derivation of the material properties required for quality assurance and for engineering design of high temperature components. However, the high temperature properties of metals and alloys are generally considered by reference to only a very limited number of parameters. Stress-rupture characteristics are normally assessed by recording the time and strain to failure whereas, even when full strain/time measurements are made, the creep behaviour is usually quantified using only one further parameter, namely the secondary or steady-state creep rate in the case of most academic studies or the time to a specified creep strain for engineering design studies. It must surely be considered debatable whether these three parameters provide an adequate description of a normal creep 'curve'. Despite their dominance of creep curve shape, relatively little attention has been focussed on either primary or tertiary creep. Yet, for example, analysis of tertiary creep behaviour in relation to the various forms of creep damage accumulation does allow a consistent approach to be formulated for the development of rejuvenative procedures for creep life enhancement of commercial creep-resistant alloys. The present overview therefore raises the question as to whether redirection of effort towards studies concerned with creep curve shape could assist in the evolution of a comprehensive theoretical treatment of creep and creep-fracture behaviour and could alleviate the problems associated with data acquisition for high temperature design.

REFERENCES

1. Parker, J. D. and Wilshire, B. (1978). *Metal Sci.*, **12**, 453.
2. Guest, J. C. (1982). In: *Measurement of High Temperature Mechanical Properties of Materials* (eds M. S. Loveday, M. F. Day and B. F. Dyson) HMSO, London, p. 23.
3. Wilshire, B., Homer, D. and Cooke, N. L. (1982). *Technological and Economic Trends in the Steel Industries*, Pineridge Press, Swansea.
4. Andrade, E. N. da C. (1910). *Proc. Roy. Soc.*, **A84**, 1.
5. Garofalo, F. (1965). *Fundamentals of Creep and Creep-Rupture in Metals*, Macmillan, New York.
6. Graham, A. and Walles, K. F. A. (1955). *J.I.S.I.*, **193**, 105.
7. Davies, P. W., Evans, W. J., Williams, K. R. and Wilshire, B. (1969). *Scripta Met.*, **3**, 671.
8. Evans, R. W., Parker, J. D. and Wilshire, B. (1982). In: *Recent Advances in Creep and Fracture of Engineering Materials and Structures* (eds B. Wilshire and D. R. J. Owen), Pineridge Press, Swansea, p. 135.
9. de Lacombe, J. (1939). *Rev. Met.*, **36**, 178.
10. Sherby, O. D. and Burke, P. M. (1967). *Prog. Mater. Sci.*, **7**, 325.
11. Monkman, F. C. and Grant, N. J. (1956). *Proc. ASTM*, **56**, 593.
12. Orr, R. L., Sherby, O. D. and Dorn, J. E. (1954). *Trans. ASM*, **46**, 113.
13. Larson, F. R. and Miller, J. (1952). *Trans. ASME*, **74**, 765.
14. Johnson, R. F. and Glen, J. (1974). *Creep Strength of Steels and High Temperature Alloys*, Metals Society, London, p. 37.
15. Raj, R. and Ashby, M. F. (1971), *Met. Trans.*, **2**, 1113.
16. Nabarro, F. R. N. (1948). *Proc. Conf. on 'Strength of Solids'*, Physical Society, London, p. 75.
17. Herring, C. (1950). *J. Appl. Phys.*, **21**, 437.
18. Coble, R. L. (1963). *J. Appl. Phys.*, **34**, 1679.
19. Ashby, M. F. (1972). *Acta Met.*, **20**, 887.
20. Evans, A. G. and Langdon, T. G. (1976). *Prog. Mater. Sci.*, **21**, 171.
21. Weertman, J. (1963). *Trans. Met. Soc. AIME*, **227**, 1475.
22. Ashby, M. F. and Frost, H. J. (1976). *Frontiers in Materials Science*, Dekker, New York, p. 391.
23. Ashby, M. F., Gandhi, C. and Taplin, D. M. R. (1979). *Acta Met.*, **27**, 699.
24. Argon, A. S. (1982). In: *Recent Advances in Creep and Fracture of Engineering Materials and Structures* (eds B. Wilshire and D. R. J. Owen), Pineridge Press, Swansea, p. 1.
25. Smith, E. and Barnby, J. T. (1967). *Metal Sci. J.*, **1**, 1.
26. Hull, D. and Rimmer, D. E. (1959). *Phil. Mag.*, **4**, 673.
27. Speight, M. V. and Harris, J. E. (1967). *Metal Sci. J.*, **1**, 83.

28. Bowring, P., Davies, P. W. and Wilshire, B. (1968). *Metal Sci. J.*, **2**, 168.
29. Ishida, Y. and McLean, D. (1967). *Metal Sci. J.*, **1**, 171.
30. Dyson, B., (1981). In: *Creep and Fracture of Engineering Materials and Structures* (eds B. Wilshire and D. R. J. Owen), Pineridge Press, Swansea, p. 235.
31. Beere, W. and Speight, M. V. (1978). *Metal Sci. J.*, **12**, 172.
32. Dyson, B. F. and McLean, D. (1972). *Metal Sci. J.*, **6**, 220.
33. Rotherham, L. A., (1951). *Creep of Metals*, Inst. Phys., London.
34. Evans, R. W. and Wilshire, B. (1981). In: *Creep and Fracture of Engineering Materials and Structures* (eds B. Wilshire and D. R. J. Owen), Pineridge Press, Swansea, p. 303.
35. Hopkin, L. M. T. (1965). *J.I.S.I.*, **203**, 583.
36. Baird, J. D. (1971). *Jern Kontorets Ann.*, **155**, 311.
37. Glen, J. (1958). *J.I.S.I.*, **190**, 114.
38. Sims, C. T. (1978). In: *High Temperature Alloys for Gas Turbines* (eds D. Coutsouradis *et al.*), Applied Science Publishers Ltd, London, p. 13.
39. Lupinc, V. (1981). In: *Creep and Fatigue in High Temperature Alloys* (ed. J. Bressers), Applied Science Publishers Ltd, London, p. 7.
40. Wilshire, B. (1978). In: *Creep of Engineering Materials* (ed. C. D. Pomeroy), Mech. Eng. Publ., London, p. 67.
41. Betteridge, W. and Heslop, J. (1974). *The Nimonic Alloys*, Edward Arnold, London.
42. White, C. H. (1981). In: *The Development of Gas Turbine Materials* (ed. G. W. Meetham), Applied Science Publishers Ltd, London, p. 89.
43. Erickson, J. S., Sullivan, C. P. and Versnyder, F. L. (1974). In: *High Temperature Materials in Gas Turbines* (eds P. R. Sahm and M. O. Speidel), Elsevier, Amsterdam, p. 315.
44. Dennison, J. P. and Wilshire, B. (1977). In: *Fourth International Conference on Fracture* (ed. D. M. R. Taplin), Univ. Waterloo Press, Vol. 2, p. 635.
45. Dennison, J. P., Holmes, P. D. and Wilshire, B. (1978). *Mater. Sci. Eng.*, **33**, 35.

4

Extrapolation and Creep Life Prediction for High Temperature Service

R. W. EVANS and B. WILSHIRE
Department of Metallurgy and Materials Technology, University College of Swansea, UK

1. INTRODUCTION

The methods adopted for design of components for elevated temperature service are usually dependent on a knowledge of the stress which the relevant materials can withstand without fracture in times up to the anticipated service life. Since these times can be 10 years or more in the case of power stations and petrochemical plant, extrapolation procedures must be used to derive the required design data. Moreover, during plant operation, there is often a need to calculate the remaining life of individual components either because the original design life has been reached or because of the wide range of temperatures which can be encountered in service.

In addition to the problems of creep life prediction and remanent life assessment, difficulties may also be posed by the need to extend the materials data base in order to take advantage of modern engineering design methods. The use of these advanced stress analysis techniques presupposes the availability of reliable long-term creep strain data rather than just stress-rupture properties. Unfortunately, relatively little information exists on the long-term creep properties of the relevant engineering materials. Further complications may then be imposed by the fact that design codes, in general, do not specifically identify welds as an important parameter, yet problems can often be encountered as a result of components in chemical and power plants not achieving their design lives because of weld failures.

In the present analysis, the limitations of current materials approaches to high temperature design are considered in relation to a new concept for data acquisition based on the development of numerical procedures for quantifying the rate of strain accumulation during creep of engineering materials.[1]

2. CONVENTIONAL PROCEDURES FOR CREEP LIFE ASSESSMENT

The criteria normally involved in the design of components subject to steady loading at elevated temperatures have been summarised[2] as:

(a) The application of the load should not cause excessive short-term yielding.
(b) Deformation during creep should not become excessive.
(c) Creep rupture should not occur during the operational lifetime.

If the first of these criteria is satisfied, then the efforts entailed in the design of components and structures for electricity generating and petrochemical plant are usually concentrated on item (b) and, more often, on item (c). Many design codes have been formulated for components of general application, such as pressure vessels and pipework. In the relevant British Standards, BS 806 and 1113, the design methods assume operation at constant known pressure and temperature, and also that the creep life of the component is related to uniaxial creep data through the mean diameter hoop stress, σ_{mdh}, as

$$\sigma_{\mathrm{mdh}} = P\frac{(d - \Delta d)}{2\,\Delta d} \qquad (1)$$

where P is the operating pressure, d is the outside diameter and Δd is the wall thickness. The necessary wall thickness can then be calculated for the operating stress by applying a safety factor (usually $\sim 1{\cdot}3$–$1{\cdot}6$) to the uniaxial stress resulting in rupture in the stipulated design life.[3]

In the case of electricity generating plant, designs are generally based on the 100 000-h rupture data for the relevant steels, although considerable economic benefits may be derived if the service lives can be extended to 250 000 h (almost 30 years). It is then necessary to devise reliable means of extrapolating short-term stress rupture data to the design lives involved because of the major costs incurred by long-term test programmes, together with the fact that periods of even 10–20 years cannot be allowed between the development of new or improved materials and their plant application.

2.1. Extrapolation of Stress-Rupture Data

Many metallurgical factors may affect the accuracy with which long-term uniaxial behaviour can be predicted from the short-term creep and stress-rupture properties of engineering alloys. The scatter encountered in short-term testing is dependent not only on the reliability of the creep equipment but also on small differences in composition within the material specification range and on minor variations in the normal heat treatment procedures employed. Even with high-precision testing facilities and precise control of the material composition and microstructure, problems are encountered because of the curvature of the log (stress)/log (time to rupture) relationships found for the types of steel widely used for the manufacture of pipework and other major components for power stations and petrochemical plant. As illustrated in Fig. 1, while short-term data up to several thousand hours indicate that a linear relationship exists between log (stress) and log (time to rupture), long-term studies[4] have demonstrated that extrapolation of short-duration tests can lead to serious overestimation of rupture life (t_f) since the stress-rupture curves deviate from linearity at longer times.

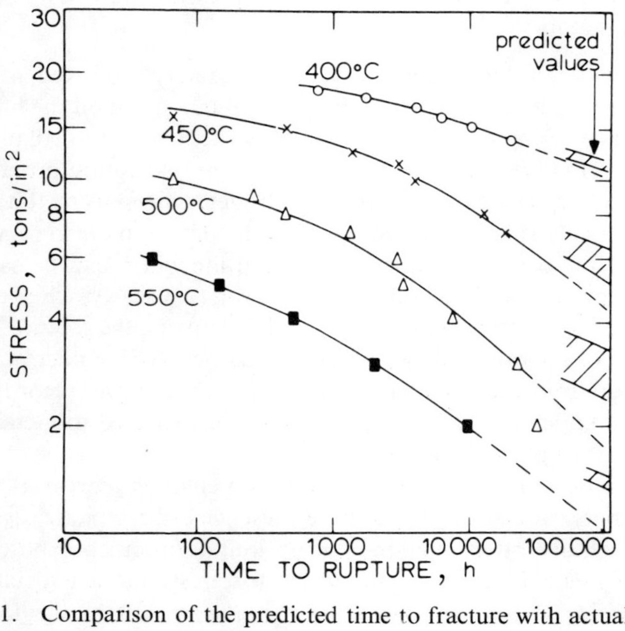

FIG. 1. Comparison of the predicted time to fracture with actual data.[4]

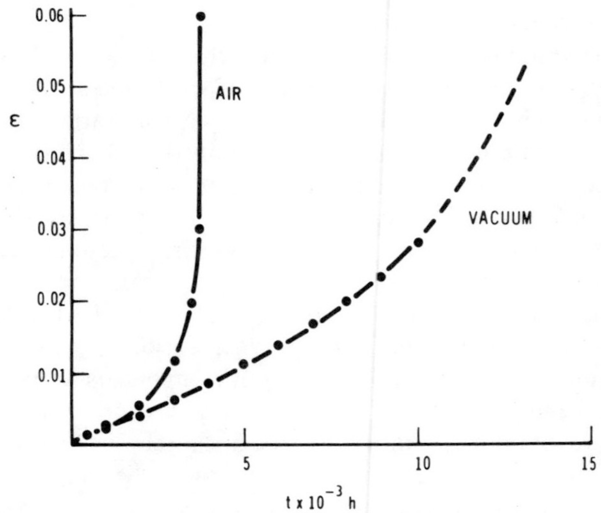

FIG. 2. Creep curves for $\frac{1}{2}$Cr–Mo–V steel tested in air and vacuum at 43 MN m^{-2} and 913 K.[5]

The curvature of the $\log(\sigma)/\log(t_f)$ relationships (Fig. 1) may be caused by several factors:

(a) Because of the large wall thicknesses usually employed, little error is introduced by ignoring the effect of oxidation on the lifetime of the low-alloy steels generally used for the construction of pipework, etc. However, the uniaxial creep and stress-rupture properties on which the designs are based must be determined using testpieces of relatively small cross-section. The distortion in creep curve shape and the resulting reduction in creep life which can be caused by testpiece oxidation is illustrated in Fig. 2. Clearly, while the early stages of the creep curve are virtually identical, the rupture life in air is considerably shorter than that observed under vacuum.[5] Moreover, the difference between the rupture lives recorded in air and under vacuum would obviously be expected to increase with increasing test duration.

(b) Commercial creep-resistant alloys generally derive their high temperature strength from the presence of precipitate dispersions which are often unstable during long-term creep exposure. The progressive loss of creep and fracture resistances associated with processes such as overageing can then result in the long-term properties being inferior to those expected on the basis of short-term tests.

(c) Problems may be encountered because the mechanisms controlling the deformation and fracture behaviour may differ in different stress and temperature regimes. Extrapolation of data from one mechanism zone to another could then overestimate the life at lower stresses.[6]

The problems involved in predicting the long-term uniaxial properties from short-term test data may therefore be dependent on the test procedures adopted, on the microstructural characteristics of the materials concerned, and also on the stress and temperature ranges over which the extrapolation procedures must be employed.

2.2. Acquisition of Long-Term Design Data

In order to establish internationally agreed property levels for inclusion in the appropriate materials standards, information provided by approved testing laboratories in Europe, America, Japan and Australia has been collated by the International Organisation for Standardisation (ISO). The creep lives recorded at different stresses and temperatures for several casts of the same material are then extrapolated by assuming that a time–temperature parameter can be specified which allows all of the stress-rupture data to be presented as a single master curve.[7] Essentially, the ISO has made use of a generalised relationship between absolute temperature and log (time), with the optional inclusion of applied stress as

$$P(\sigma) = \frac{(\sigma^{-q} \log t) - \log t_a}{(T - T_a)^r} \tag{2}$$

where $P(\sigma)$ is the parameter value, σ is the applied stress, t is the testing time, T is the absolute test temperature, and q, t_a, T_a and r are selected constants.[8] This generalised equation can then be reduced to specific forms. For example, when $q = 0$ and $r = 1$ eqn (2) leads to the Manson–Haferd equation[9]

$$P(\sigma) = \frac{\log t - \log t_a}{T - T_a} \tag{3}$$

Similarly, when $q = 0$, $r = -1$ and $T_a = 0$

$$P(\sigma) = T(\log t - \log t_a) \tag{4}$$

which is the Larson–Miller relationship.[10]

As a result of the uncertainties associated with procedures involving the use of arbitrary constants, extrapolation is generally limited to three times the longest test duration available. Thus only when results are available to, say, 30 000 h are the 100 000-h predictions reasonably accurate.[4] Furthermore, the scatter in the multi-laboratory data is quoted as $\pm 20\%$ of the stress which is equivalent to about an order of magnitude scatter in rupture lives at any specified stress level.[7] Since the lower limit of the scatter band is usually used for design purposes, the procedures adopted may be unnecessarily conservative. An additional problem is then introduced by the need to extend the materials data base in order to take full advantage of modern design procedures.[11] In particular, as improved stress analysis techniques become available, reliable long-term creep strain data become vital. Unfortunately, relatively little information is available in this area. Consequently, attention is now being focussed on the correlation of creep and rupture strengths which allows the stress to produce a given strain in a specified time to be estimated. However, this procedure results in a scatter band which is slightly greater even than the available stress rupture data.[11]

2.3 Remanent Life Assessment

Components and structures for power stations and petrochemical plant are usually designed not on the concept of an ultimate life but merely on the basis that failure will not occur within the design life under the operating conditions anticipated. The rarity of serious failures can then be taken as evidence for the conservatism of current design codes which introduce a safety factor in addition to the assumption of minimum materials performance (i.e. the lower limit of the scatter band associated with the relevant stress-rupture data). However, during the last decade, economic pressures have provided an incentive to extend the operational life of existing plant. Procedures have therefore been sought for estimating the remanent life of components,[12] either because the original design life has been reached or because of uncertainties over the actual operating conditions encountered.

One procedure for estimating remanent life entails monitoring the service conditions so that future life expectancy can be assessed by comparing, at a corresponding stress and temperature, the total period of service with the lower limit of the scatter band of the stress-rupture data for the relevant material. However, this approach is unlikely to offer high accuracy because of the difficulties of monitoring service conditions, the problems of deciding upon the uniaxial stress which is 'equivalent' to that experienced in service, and the uncertainties introduced by the large scatter bands associated with the stress-rupture data. A more accurate assessment of life usage may then be based on measurements of the rate of strain accumulation on operational plant. Measurements may be taken either of the external dimensions of a component or of localised strain patterns but, as with most procedures for remanent life assessment, difficulties will be posed by the inhomogeneous nature of strain accumulation with complex component geometries.

The most widely adopted approach involves measurement of some material characteristic after various periods of service in an attempt to quantify creep damage. One such procedure applies what is usually termed the 'life-fraction' or Robinson's rule.[13] Essentially, this approach assumes that, for two creep exposure conditions (s and a) giving rupture lives of t_{fs} and t_{fa}, respectively (with $t_{fs} > t_{fa}$), after a period t_s under condition 's' the further life expected on changing to condition 'a' is t_a when

$$\frac{t_a}{t_{fa}} + \frac{t_s}{t_{fs}} = 1$$

or for Z changes of exposure conditions

$$\sum_{i=1}^{Z} \frac{t_i}{t_{fi}} = 1 \tag{5}$$

Samples taken from operational components after known periods of service may then be taken to failure under accelerated test conditions so that the remanent service life can be estimated. Unfortunately, a comprehensive pattern of material behaviour has not yet been established for post-exposure testing. For low-alloy steels in particular, it appears that the life-fraction rule appears more satisfactory when temperature increases rather than stress increases are chosen as the means of providing accelerated test conditions.[12]

Post-exposure testing based on Robinson's rule or other similar procedures often requires a number of samples of a size sufficient to allow testpieces of the requisite dimensions to be produced. If sampling of service components does cause problems, a suitable alternative may be offered by compression testing which would satisfactorily identify the loss of creep strength associated with microstructural instability, with the advantage that high-precision data could be obtained using only very small testpieces.[14] Similarly, small samples are adequate for microstructural studies which estimate remanent life through measurement of the extent of precipitate coarsening, changes in precipitate composition, fracture damage, etc.[15–17] Indeed, microstructural studies involving only surface examination of components appear to provide a valuable guide to life expectancy.[18] This approach is limited only by the need for access but has the advantage of being equally suitable for inspection of regions where fracture is likely to originate, namely welds. In this way, since creep damage can be revealed at an early stage, components can generally be monitored and operated up to the point at which repair becomes mandatory.[18]

3. THE θ PROJECTION CONCEPT

A knowledge of the stress which a material can withstand without fracture for any pre-selected time is essential not only for the design of high temperature components but also for assessment of the remanent life of operational plant. In order that all available stress analysis procedures can then be employed to design plant with a degree of conservatism sufficient only to ensure safety, it is also necessary to provide not just accurate stress-rupture properties but also reliable long-term creep strain data for the relevant materials over the stress and temperature conditions expected in service. In addition, it would be highly desirable if the necessary creep and creep-rupture data spanning the time-scales up to the required design lives could be obtained without the penalties and uncertainties associated with long-term test programmes. The present section introduces just such a procedure[1] which, if fully validated, will provide the means of extrapolating high-precision short-term uniaxial creep results to predict the full data range needed for high temperature design.

3.1. Representation of Creep Curve Shape

In order that the properties forecast by this new extrapolation procedure could be assessed in relation to currently available long-term data, a study was made of the creep behaviour of $\frac{1}{2}$Cr–$\frac{1}{2}$Mo–$\frac{1}{4}$V ferritic steel,[1] a material widely used in British electricity generating plant. Using high-precision constant-stress machines, creep tests were carried out at 838 K which is the design temperature for components produced from low-alloy steels for power station service. Over a range of stresses from 125 to 306 MN m^{-2}, normal creep curves were invariably recorded. Following the initial specimen strain on loading, the creep rate decreased continuously during the primary stage until a minimum rate was attained before the creep rate eventually accelerated during the tertiary stage which preceded fracture. In all cases, fracture occurred in an intergranular manner. However, the appearance of the true strain/time curves changed considerably over the stress range examined (Fig. 3). The distinct primary stages evident at stresses of $\sim 300\ \mathrm{MN\,m^{-2}}$ became less pronounced as the total primary creep strain decreased with decreasing applied stress. In contrast, the tertiary stage began progressively to dominate the form of the creep curve with decreasing stress such that, for stresses of $\sim 200\ \mathrm{MN\,m^{-2}}$ and below, the primary stage could be resolved only by detailed examination of the precise strain/time record obtained early in the creep life (Fig. 3).

For each test condition it was found that the variation in creep strain, ε, with time, t, could be described accurately as[1]

$$\varepsilon = f\{\theta, t\} = \theta_1[1 - \exp(-\theta_2 t)] + \theta_3[\exp(\theta_4 t) - 1] \tag{6}$$

where $\theta = [\theta_1, \theta_2, \theta_3, \theta_4]^{\mathrm{T}}$. This equation sees the creep curve as an addition of decaying and accelerating exponential functions of time in which θ plays the role of governing creep curve shape. This can be most readily seen by differentiating eqn (6) to give

$$\frac{d\varepsilon}{dt} = \theta_1\theta_2 \exp(-\theta_2 t) + \theta_3\theta_4 \exp(\theta_4 t) \tag{7}$$

and

$$\frac{d^2\varepsilon}{dt^2} = -\theta_1\theta_2^2 \exp(-\theta_2 t) + \theta_3\theta_4^2 \exp(\theta_4 t) \tag{8}$$

θ_1 and θ_3 act as scaling parameters which control the extent, with respect to strain, of the primary and tertiary stages of creep. θ_2 and θ_4 have an important effect on curve shape and eqn (8) shows that they sensitively affect the curvature of the primary and tertiary stages, i.e. increasing θ_2 and θ_4 rapidly increases the deceleration of creep rate in the primary stage and the acceleration in the tertiary stage.

For each creep test, between 100 and 200 strain/time readings were recorded which allowed the θ parameters to be determined precisely. The numerical procedures, the full computer listings and the operating instructions necessary have been detailed elsewhere.[1] Using these procedures, it was established that eqn (6) provides an accurate description of the rate of creep strain accumulation at each stress level considered. Moreover, the gradual change in creep curve shape with increasing test

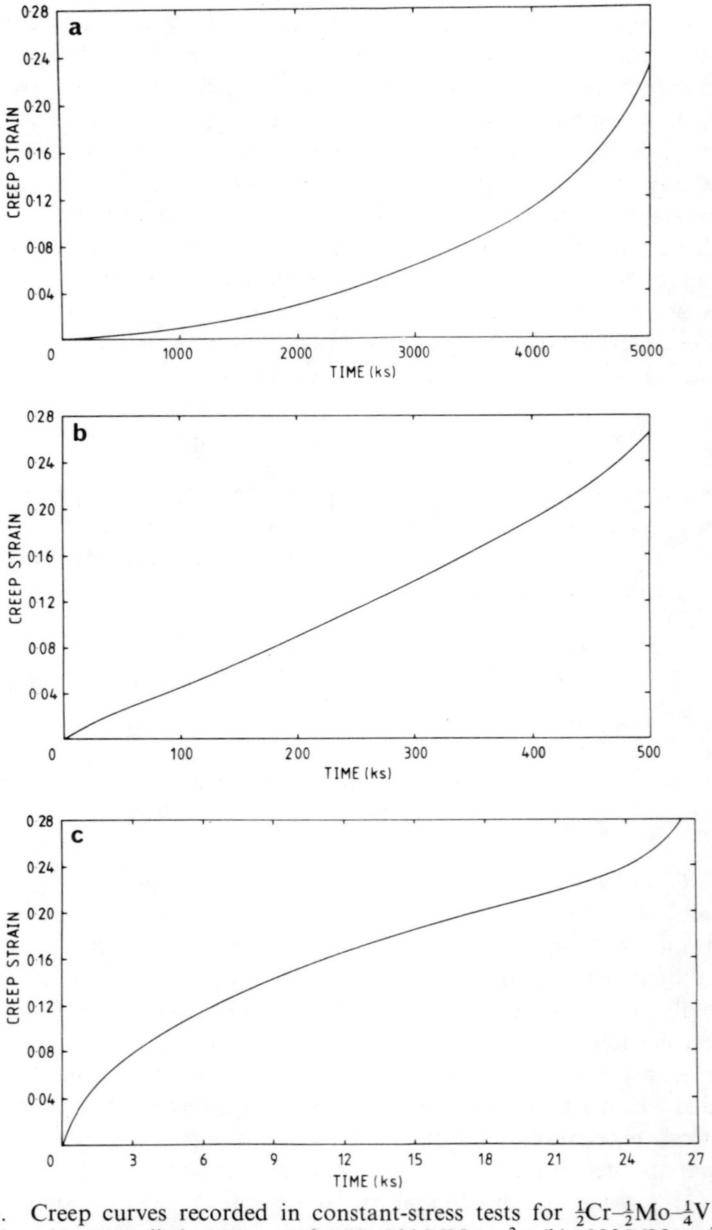

FIG. 3. Creep curves recorded in constant-stress tests for $\frac{1}{2}$Cr–$\frac{1}{2}$Mo–$\frac{1}{4}$V steel carried out at applied stresses of: (a) 180 MN m^{-2}, (b) 223 MN m^{-2}, (c) 278 MN m^{-2} at 838 K.[1]

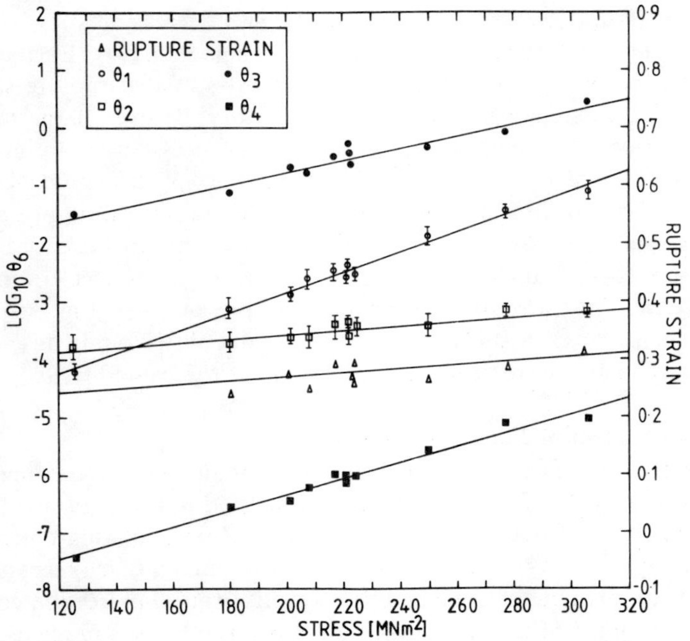

FIG. 4. Variation of θ and rupture strains with stress at 838 K. Error bars for θ_1 and θ_2 are 95% confidence intervals. The confidence intervals for θ_3 and θ_4 are too small to be represented on the present scale.[1]

duration can be quantified precisely in terms of the stress dependence of θ (Fig. 4) as

$$\theta_1 = \exp(-14{\cdot}5 + 0{\cdot}040\sigma)$$

$$\theta_2 = \exp(-9{\cdot}85 + 0{\cdot}0081\sigma)$$

$$\theta_3 = \exp(-6{\cdot}51 + 0{\cdot}024\sigma)$$

$$\theta_4 = \exp(-21{\cdot}03 + 0{\cdot}032\sigma) \qquad (9)$$

While eqns (6) and (9) appear to offer a highly satisfactory representation of creep curve shape, this type of analysis does not specifically refer to fracture. In order to complete the description of material behaviour, it is necessary to define some criterion of failure. Clearly, for any creep curve, this may be done by specifying either the rupture time or the rupture strain. Since only one of these two quantities is needed, it is obviously desirable to use the quantity which shows the smallest variation with stress. With $\frac{1}{2}$Cr–$\frac{1}{2}$Mo–$\frac{1}{4}$V steel, the rupture life varies markedly with stress whereas the rupture strain decreases only very slowly with decreasing stress over the entire range of test conditions examined (Figs 3 and 4). Furthermore, defining fracture as the point at which an extrapolated rupture strain is reached causes relatively little error in rupture life even if the material ductility does change unexpectedly at very low stress levels. In low stress tests the creep curves are dominated by the tertiary processes, defined by θ_3 and θ_4. The strain/time behaviour may then be considered simply in terms of a gradually increasing creep rate throughout virtually the entire life. In the late stages of the tertiary process, the creep rate is accelerating very rapidly, as illustrated by the creep curve presented in Fig. 3(a). As a consequence, even a substantial change in ductility would cause only a relatively small variation in rupture life under these conditions.

3.2. Assessment of Long-Term Predictions

The linearity of the stress/log θ relationships and the narrow confidence intervals on each θ function offer a new method not only of quantifying the changes in creep curve shape but also of extrapolating creep data. Essentially, for any specified stress, eqns (6) and (9) may be used to construct a predicted creep curve. However, if this 'θ projection concept' is to be adopted, it is important to establish whether the predicted creep properties agree reasonably well with actual long-term behaviour.

A check on the accuracy of the 'θ projection concept' is provided by comparisons of the predicted and measured values of the minimum creep rate.[1] The tertiary stage begins at a progressively earlier fraction of the total creep life with decreasing applied stress (Fig. 3). Consequently, the minimum creep rate can be measured after a relatively short time even in low stress tests and, since the minimum rate is attained after only very low creep strains, the measured value will be independent of whether constant-stress or constant-load equipment is used. Moreover, the theoretical value of the minimum creep rate can be derived easily for any applied stress level. Equation (8) indicates that the function $f\{\theta, t\}$ has a minimum gradient for positive values of θ_i and that this minimum creep rate occurs at time t_m given by

$$t_m = \frac{1}{\theta_2 + \theta_4} \ln \frac{\theta_1 \theta_2^2}{\theta_3 \theta_4^2} \qquad (10)$$

Thus, if θ is known at any set of testing conditions, it is possible to calculate t_m and hence, through eqn (7), the minimum creep rate, $\dot{\varepsilon}_m$. The predicted stress dependence of the minimum creep rate is shown in Fig. 5 for stresses down to those encountered during operation of power stations. Clearly, the

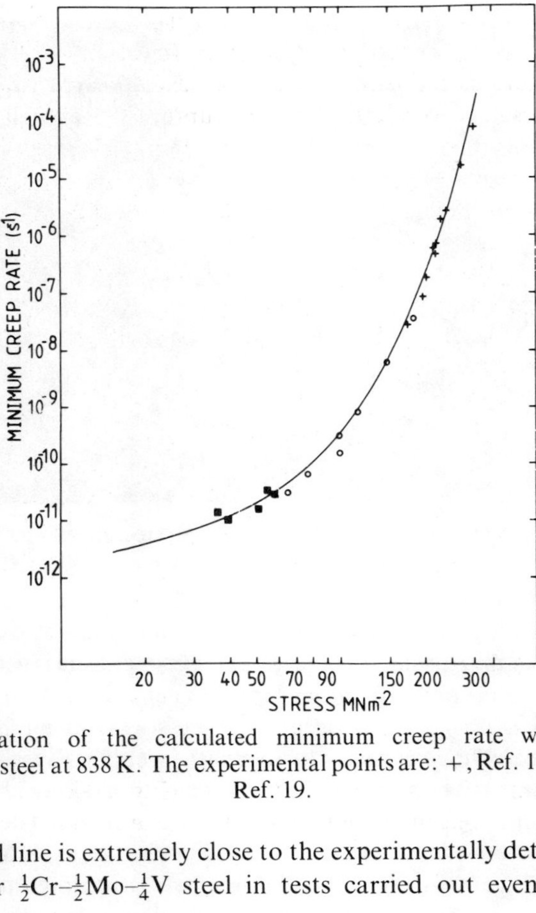

FIG. 5. Variation of the calculated minimum creep rate with stress for $\frac{1}{2}$Cr–$\frac{1}{2}$Mo–$\frac{1}{4}$V steel at 838 K. The experimental points are: +, Ref. 1; ○, Ref. 3; ■, Ref. 19.

extrapolated line is extremely close to the experimentally determined rates reported for $\frac{1}{2}$Cr–$\frac{1}{2}$Mo–$\frac{1}{4}$V steel in tests carried out even at very low stresses.[3,19]

3.3. Implications of the θ Projection Concept

Apart from providing a source of confidence in the θ projection concept, the close correlation between the predicted and the measured minimum creep rates (Fig. 5) raises issues fundamental to conventional interpretations of creep behaviour.

Most studies of high temperature creep behaviour discuss the stress dependence of the 'secondary' creep rate in terms of a power law representation (i.e. $\dot{\varepsilon}_s \propto \sigma^n$). It is then common to represent the $\log \sigma / \log \dot{\varepsilon}_s$ relationships by means of straight-line segments of gradient n. With $\frac{1}{2}$Cr–$\frac{1}{2}$Mo–$\frac{1}{4}$V steel (Fig. 5) this approach would result in a decrease in n

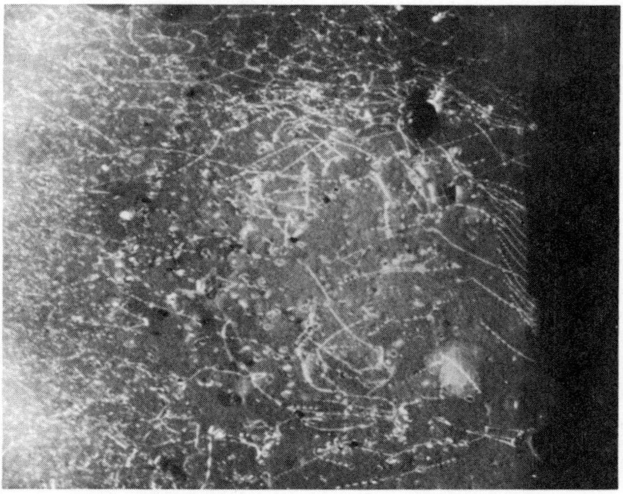

FIG. 6. Microstructure developed in $\frac{1}{2}$Cr–$\frac{1}{2}$Mo–$\frac{1}{4}$V steel after 50 000 h of service at $\sim$40 MN m^{-2} and $\sim$838 K in a reheat-header ($\times$42 000 dark field).

value from 4 at high stress levels to almost unity at low stress levels. This change in the magnitude of the stress exponent has generally been interpreted in terms of a transition in creep mechanism. With n values of 3 or more, creep is considered to occur by processes involving the generation and movement of dislocations. In contrast, n values close to unity are normally assumed to be associated with deformation taking place by stress-directed flow of vacancies without dislocation movement. The relevance of dislocation processes in the $n \simeq 1$ regime with $\frac{1}{2}$Cr–$\frac{1}{2}$Mo–$\frac{1}{4}$V steel is illustrated in Fig. 6, which shows the microstructure present after long-term exposure in electricity generating plant. Similar dislocation arrangements are observed both in long-term creep tests carried out at 40 MN m^{-2} at 838 K and in high stress tests lasting less than 1000 h at the same temperature.[20] These observations indicate that creep deformation of low-alloy steels occurs by the generation and movement of dislocations even in the stress range where $n \simeq 1$.

A direct interpretation for the curvature of the stress/minimum creep rate relationship for $\frac{1}{2}$Cr–$\frac{1}{2}$Mo–$\frac{1}{4}$V steel can, however, be provided by the analysis of creep curve shape based on eqn (6). Unlike conventional approaches which discuss the changes in the gradient of the $\log \sigma/\log \dot{\varepsilon}_s$ plots in terms of transitions in creep mechanism, no change in creep process need be invoked to explain this effect using the θ projection concept. Under

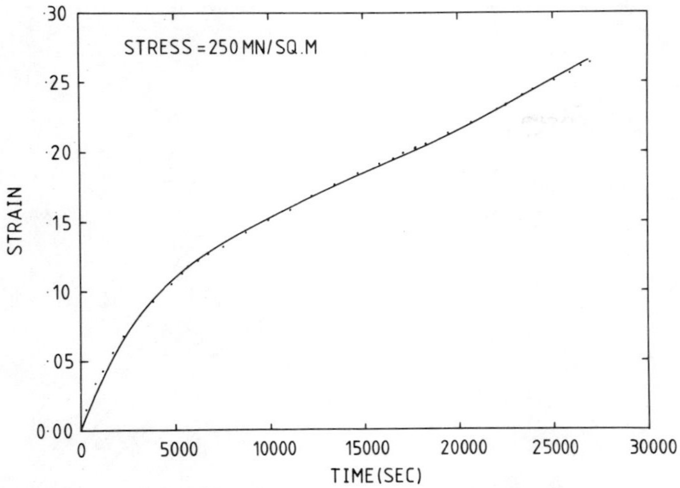

FIG. 7. Creep curve recorded for $\frac{1}{2}$Cr–$\frac{1}{2}$Mo–$\frac{1}{4}$V steel in a constant-stress test carried out at 250 MN m^{-2} and 838 K, showing the experimental strain/time readings in relation to the calculated curve. For clarity, less than one in ten of the actual strain/time readings taken are presented.

all conditions examined, the creep curve shape could be described accurately using eqn (6) (Fig. 7) and the changing form of the curves with increasing test duration (Fig. 3) could be quantified by linear stress/log θ functions (eqn (9)). Yet, while the stress/log θ plots are linear (Fig. 4), the complex dependence of the minimum creep rate on θ causes the stress/minimum creep rate relationship to curve markedly with decreasing stress (Fig. 5).

Analytical procedures based on eqn (6) then imply that the minimum creep rate, although the simplest of creep measurements, cannot be taken as a parameter which adequately 'characterises' creep properties. Instead, presentation of creep data in the form of iso-strain graphs showing the times to attain various creep strains as a function of stress at any temperature (Fig. 8), which can be related to the θ functions (eqn (9)), would appear to be of greater fundamental significance than standard descriptions of creep data which specify only the minimum creep rate.

For the iso-strain data presented in Fig. 8, the time to reach a total strain of 0·30 may be approximated to the rupture life of the $\frac{1}{2}$Cr–$\frac{1}{2}$Mo–$\frac{1}{4}$V steel at 838 K. Moreover, as with the log σ/log $\dot{\varepsilon}_m$ data shown in Fig. 5, the curvature of this log σ/log t_f relationship can be explained in a manner which is fully consistent with the microstructural evidence reported for

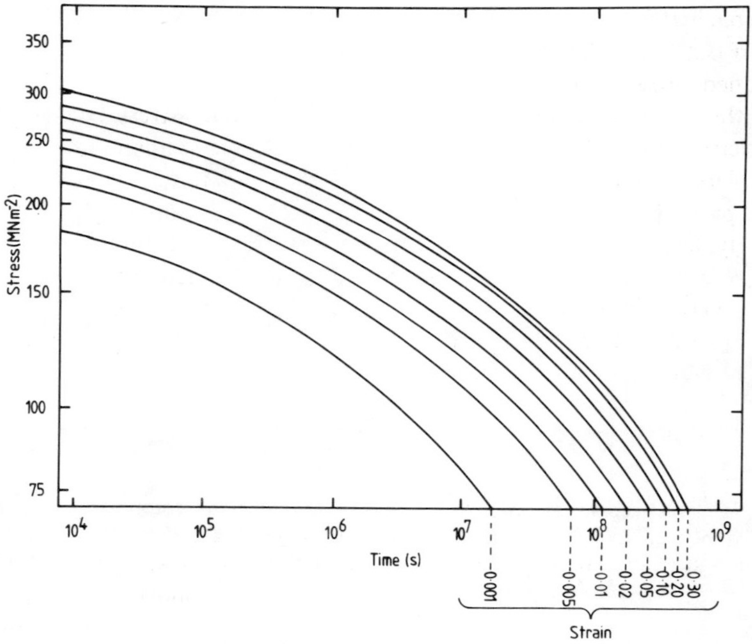

FIG. 8. Calculated iso-strain data for $\frac{1}{2}$Cr–$\frac{1}{2}$Mo–$\frac{1}{4}$V ferritic steel at 838 K. The curves are for constant true strains between 0·001 and 0·3 under constant-stress conditions.[1]

$\frac{1}{2}$Cr–$\frac{1}{2}$Mo–$\frac{1}{4}$V steel.[1] At all stress levels, fracture eventually occurs in an intergranular manner as grain-boundary cavities and cracks develop predominantly when necking of the specimen becomes apparent late in the creep life. However, except at very high stresses when tertiary creep begins because of necking, the commencement of the tertiary stage in long-duration tests is attributable to the progressive loss of creep strength accompanying coarsening of the carbide dispersion present. Since the effects of overageing become more pronounced as the test duration increases, the tertiary stage increasingly dominates the creep curve shape with decreasing stress (Fig. 3). Analysis of strain/time behaviour using eqn (6) suggests that the minimum creep rate and the rupture life are variables which are entirely dependent on the shape of the creep curve and the rupture strain. On this basis, it is proposed that the serious overestimation of long-term performance which results from conventional extrapolations of $\log \sigma/\log \dot{\varepsilon}_m$ and $\log \sigma/\log t_f$ plots constructed from short-term data can be avoided by extrapolating the θ parameters. With this new approach, the minimum creep rate and the rupture life as well as the full iso-strain data may then be derived using the appropriate θ values for any specified stress level.

Although the maximum test duration of the constant-stress data used for the derivation of the θ functions (eqn (9)) was less than three months, the complete behaviour patterns for $\frac{1}{2}$Cr–$\frac{1}{2}$Mo–$\frac{1}{4}$V ferritic steel at 838 K have been predicted for creep lives of up to 10 years and more (Figs 5 and 8). Clearly, full validation of the θ projection concept will require comparison with creep strain and creep life data being obtained by laboratories concerned with the acquisition of long-term data. However, if this new extrapolation procedure can be validated, numerous technical and cost advantages follow automatically:

(a) While the costs of the high-precision constant-stress equipment required for the θ determinations are greater than those for conventional constant-load machines, and particularly for standard constant-load stress-rupture units, these costs would be trivial in relation to the short time-scales which would be needed to estimate long-term behaviour. Moreover, the short-term nature of the basic information needed for extrapolation would eliminate problems associated with testpiece oxidation during long-term programmes and obviate the need for expensive vacuum or controlled-atmosphere testing.

(b) The θ projection concept offers 'constant-stress' data rather than the less useful 'constant-load' results usually provided. Constant-stress conditions mean that the load on the specimen is adjusted continuously to compensate for the uniform decrease in cross-sectional area as the testpiece deforms. With constant-load tensile creep machines, the stress quoted is the initial stress on the sample at the commencement of the test, but the actual stress on the material increases with increasing creep strain. This effect can lead to a serious distortion of creep curve shape for materials exhibiting significant strains to fracture, i.e. elongations to failure of more than a few percent.

(c) The distortion of creep curve shape and the consequent reduction in creep life associated with constant-load test procedures may be the factors responsible for the failure of Robinson's rule to give reliable estimates of remanent life when 'stress-accelerated' tests are performed on samples of ferritic steels after service exposure. For this reason, a major research programme is now underway in

Swansea, sponsored by the Science and Engineering Research Council in association with the Central Electricity Generating Board. This programme aims to determine whether the 'θ projection concept' can be extended to provide accurate remanent life procedures based on post-exposure testing of austenitic and ferritic steels.

(d) The θ projection concept predicts not only minimum creep rate and rupture life but also full details of the rate of strain accumulation for any specified test conditions. While reliable iso-strain data are necessary for use with any design code which defines the useful service life as the time to attain a particular creep strain (say 1 %), such detailed information as the variation of creep strain rate at all creep strains is required as input parameters when finite element methods are used to calculate the stresses and strains in components operating at high temperatures. If fully validated, the θ projection concept will then provide the complete data base needed if advantage is to be taken of modern stress analysis procedures for high temperature design.

4. CONSTITUTIVE EQUATIONS

The design of components for high temperature service is complicated by the strongly non-linear creep behaviour of metals and alloys. Analytical approaches can yield useful results for simple geometric shapes and reasonably steady loadings but the current availability of large fast computing facilities has allowed analysis to be considerably extended. Finite difference and finite element procedures permit quite general shapes to be investigated and also make possible the inclusion of creep properties which vary from point to point. Such design investigations can be carried out incrementally with suitable time-stepping routines allowing the gradual change of component shape to be calculated, large strains to be evaluated, and sharp changes in temperature and loading to be taken into account. The necessary numerical techniques are well established and widely available but the results they yield can only be as good as the material behaviour descriptions which are embodied in the constitutive equations. These properties are often poorly understood and this frequently gives rise to a major source of uncertainty.

Some constitutive relationships are expressed in terms of the creep strains which might accumulate with time as a function of stress and temperature but the numerical techniques now available more usually require a relationship which relates creep strain rate to some function of the prevailing stress state. The use of such a relationship allows the calculation of the velocities of all points in the creeping body and a suitable time integration scheme can calculate geometry changes. In addition, a full constitutive equation needs to contain some description of fracture behaviour so that the time and position at which creep fracture occurs can be estimated.

4.1. Uniaxial Creep Behaviour

The majority of experimental creep data is collected by means of uniaxial tensile tests so that it is not surprising that the development of constitutive equations has concentrated on this simple stress state. The creep strain rate ($\dot{\varepsilon}$) is related to stress (σ) and temperature (T) by either

$$\dot{\varepsilon} = f_1(\sigma, T, t) \tag{11a}$$

or

$$\dot{\varepsilon} = f_2(\sigma, T, \varepsilon) \tag{11b}$$

Equation (11a) relates the creep rate to the expired time (t) whereas eqn (11b) expresses it in terms of the current creep strain (ε). It is important that the functions chosen are capable of representing the full creep curve. Many heat-resisting steels and commercially used superalloys have creep curves which are heavily dominated by the tertiary stage so that descriptions which are confined to primary and secondary creep are not adequate. One method of simplifying eqns (11a) and (b) is to assume that they can be split into separate functions of the variables, namely

$$\dot{\varepsilon} = u_1(\sigma)v_1(T)w_1(t) \tag{12a}$$

or

$$\dot{\varepsilon} = u_2(\sigma)v_2(T)w_2(\varepsilon) \tag{12b}$$

In such cases, the various functions can, in principle, be determined from simple isothermal constant-stress creep curves. Various possibilities have been suggested for the form of these functions. The stress function may be either a simple power relationship or an exponential, or a sinh term which tends to the two simpler forms at low and high stresses, respectively. The temperature function is often taken to be an Arrhenius type exponential. For the time (or strain) function, the relationship has to be capable of representing primary and tertiary creep, and one such relationship is that given by Graham and Walles.[21] In this case, a suitable uniaxial description of the creep rate is

$$\dot{\varepsilon} = A\sigma^n \exp\left(-\frac{Q}{RT}\right)(a + bt^{-2/3} + ct^2) \tag{13}$$

where A, n, a, b and c are experimentally determined constants. The strain form of this equation (see eqn (12b)) can be readily found by substituting into eqn (13) the time $t = t'$, where t' is the solution of the equation

$$3bt^{1/3} + at + \tfrac{1}{3}ct^3 - \varepsilon = 0 \tag{14}$$

In practice, the separation of variables implied in eqn (13), although mathematically convenient, can rarely be experimentally justified. Thus the experimentally determined constants are frequently themselves functions of stress and temperature. It is probable that the best approach is to abandon the separation since a simpler overall expression may result. One such possibility is the use of eqn (7) as a constitutive equation written in terms of a time parameter. The corresponding strain parameter relationship can be obtained by substituting into eqn (7) the time $t = t'$ given by the (numerical) solution of the equation

$$\theta_1[1 - \exp(-\theta_2 t)] + \theta_3[\exp(\theta_4 t) - 1] - \varepsilon = 0 \tag{15}$$

The variation of creep rate with stress and temperature is now accommodated by making the various θ_i functions of these variables. For a number of materials, linear functions suffice so that the expressions

$$\ln \theta_i = a_i + b_i T + c_i \sigma + d_i \sigma T \qquad (i = 1, 4) \tag{16}$$

together with eqn (7), provide the basis of a good constitutive equation. The determination of the various constants will require high-accuracy creep curves obtained under isothermal constant-stress conditions.[1] It is because of this lack of adequate basic creep data (as well as for mathematical simplicity) than many very approximate constitutive equations are used. At the extreme, analyses are often conducted for isothermal conditions with b and c of eqn (13) put equal to zero. In this case, the relationship reduces to Norton's law

$$\dot{\varepsilon} = A\sigma^n \tag{17}$$

so that it is no longer necessary to distinguish between the strain and time parameter forms of eqn (12).

Creep rupture under uniaxial conditions is a strain-controlled phenomenon and it is possible that the stipulation of a maximum attainable strain is sometimes a sufficient fracture criterion. It is also possible to develop a fracture constitutive equation in terms of a damage parameter, ω, which will vary between 0 and 1 during the full creep life. The rate of change of this

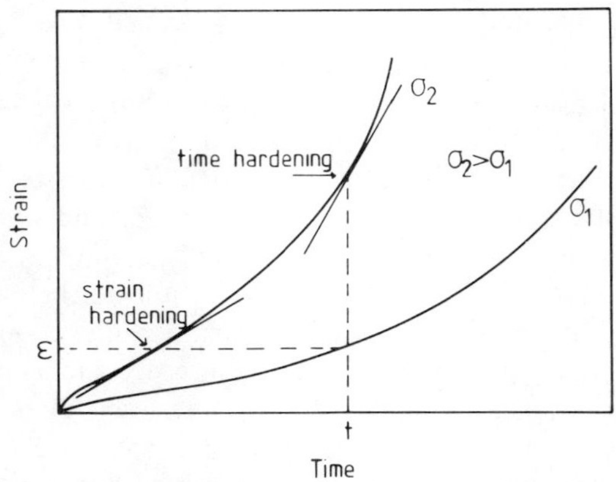

FIG. 9. Strain-hardening and time-hardening rules for creep rate estimation following a stress change from σ_1 to σ_2.

parameter can be related to the current state of stress through an equation similar to Norton's law

$$\dot{\omega} \propto \sigma^k \tag{18}$$

so that, if increases in ω are interpreted as reductions in the stress-bearing area of the component, the necessary integrations can be performed to yield an estimate of rupture life.

4.2. Creep Under Non-Steady Conditions

In many service environments, the stress and temperature conditions can change with time, often sharply. It is essential that any design model can deal with these changes, so that the new creep rates associated with the new conditions can be calculated. One procedure is simply to use eqns (11a) and (b), with updated stresses and temperatures. If eqn (11a) is used, this is called a time-hardening procedure whereas eqn (11b) leads to a strain-hardening method. The physical meaning of these terms is illustrated in Fig. 9, which shows two isothermal creep curves at stresses σ_1 and σ_2 ($\sigma_2 > \sigma_1$). Creep proceeds at stress σ_1 until, at some time t (strain ε), the stress is increased abruptly to σ_2. The figure illustrates that different estimates of the new creep rate will be obtained by the two hardening methods. There appears to be no good physical reason for preferring either estimate but there is some experimental evidence that strain hardening may be slightly more accurate. One of the conveniences of using the simplified Norton's law is that the choice of hardening process is avoided since they yield the same result.

This problem of creep transient behaviour has led to the development of more complex types of constitutive equation. It is assumed that creep rate is not only a function of stress and temperature but also of one or more 'internal variables' which are themselves functions of the history of the material. The internal variables are intended to describe the nature of metallurgical structural changes (such as recovery and work hardening) and hence allow realistic estimates of transient behaviour. A wide variety of models are available,[22] the more complex requiring the evaluation of a large number of material constants. Furthermore, these constants cannot be obtained from simple creep tests but require stress relaxation, constant strain rate, and stress and temperature cycling tests. Thus, although the concepts are likely to lead to more precise constitutive equations, few engineering materials have yet been sufficiently well investigated to allow detailed evaluation to be attempted.

4.3. Creep Under Multiaxial Stresses

Multiaxial stress states have been investigated by the use of biaxial plates, torsion specimens and pressurised thin-walled tubes. Experimentation suggests that creep deformation is incompressible and that the presence of hydrostatic pressures has no effect on the components of the strain-rate tensor. Moreover, these strain-rate components are proportional to the deviatoric stress tensor components. This suggests that the Levy–Mises equations form a convenient basis for a multiaxial constitutive relationship. Thus

$$\dot{\varepsilon}_{ij} = \lambda S_{ij} \tag{19}$$

where $S_{ij} = \sigma_{ij} - (\sigma_{11} + \sigma_{22} + \sigma_{33})/3$, and λ is a constant of proportionality. The von Mises yield criterion allows λ to be determined so that

$$\dot{\varepsilon}_{ij} = \frac{3\bar{\dot{\varepsilon}}}{2\bar{\sigma}} S_{ij} \tag{20}$$

where the effective strain rate ($\bar{\dot{\varepsilon}}$) and effective stress ($\bar{\sigma}$) can be defined in terms of the principal strain rate and stress components as

$$(\sigma)^2 = \tfrac{1}{2}[(\sigma_1 - \sigma_2)^2 + (\sigma_2 - \sigma_3)^2 + (\sigma_3 - \sigma_1)^2] \tag{21}$$

and

$$(\bar{\dot{\varepsilon}})^2 = \tfrac{2}{9}[(\dot{\varepsilon}_1 - \dot{\varepsilon}_2)^2 + (\dot{\varepsilon}_2 - \dot{\varepsilon}_3)^2 + (\dot{\varepsilon}_3 - \dot{\varepsilon}_1)^2] \tag{22}$$

The numerical constants in these expressions are chosen so that the effective stress and strain rate are equal to the uniaxial values for the simple tensile test. Experiments have also shown that, to a good approximation, the relationship between effective stress and effective strain rate under complex conditions is the same as that between stress and strain rate for uniaxial testing so that

$$\bar{\dot{\varepsilon}} = f_1(\bar{\sigma}, T, t) \quad (23a)$$

or

$$\bar{\dot{\varepsilon}} = f_2(\bar{\sigma}, T, \bar{\varepsilon}) \quad (23b)$$

Combining eqns (20) and (23) yields a simple constitutive relationship which utilises all the data available for the equations of Section 4.1.

Transient behaviour can be estimated by the strain- or time-hardening rules as for the uniaxial case. More general internal variable models are also available, but the large number of material constants required to deal with the multiaxial stress state make the determination of suitable relationships very difficult.

It appears that the multiaxial creep fracture behaviour of metals cannot be immediately derived from the uniaxial rules in the same way as the creep-rate relationships. The rupture life does not depend solely on the effective stress, but on the maximum principal stress or the ratio of maximum principal stress to effective stress.[23] The extension of damage parameter concepts to multiaxial stress states is thus not straightforward.

5. PREDICTION OF WELD PERFORMANCE

Welds are an essential feature of large-scale plant for pressure containment as, for example, in chemical installations, the oil and gas industries, and in power generation. Indeed, the performance of welds is often the life-limiting factor in many components and structures, particularly under service conditions where creep failure can occur. Yet, in general, design codes do not specifically identify welds as an important parameter. In design terms, welds are normally accounted for simply by postulating a 'weld-efficiency factor'. In many cases this merely involves down-rating the allowed stress calculated for an homogeneous component and by stipulating that the ductility of any part of the weld should be adequate for the application envisaged.

Even with high-quality welds, there are special features of the weldments which are not encountered with homogeneous components. In particular, the microstructures and therefore the properties of steels are modified in the heat affected zone (HAZ) formed in regions of the parent steel adjacent to the weld. Assessment of weld performance is then complicated not only by the variety of microstructures and properties involved but also by the wide range of weld geometries usually found in practice.

On this basis, analysis of weld performance in high temperature plant is considerably more complex even than that of predicting the long-term behaviour of homogeneous components. However, since welds will remain a major feature of large-scale plant, it is essential that an adequate solution to this seemingly intractable problem is eventually found. Essentially, the failure characteristics of weldments under high temperature creep conditions have been approached by large-scale testing of individual components or by laboratory testing of suitably designed 'cross-weld' testpieces. Unfortunately, while the latter procedure is attractive in terms of time and cost, its use has been inhibited by the lack of any knowledge as to the specimen geometries and test procedures which provide the most meaningful creep data with respect to the weld.

A comprehensive review of the influence of welding on the creep properties of steels has been published recently.[24] Rather than attempting to duplicate this review, the present section has therefore been designed to provide an illustration of the way in which high-precision cross-weld testing combined with finite element analysis may be used to account quantitatively for a specific type of weld failure frequently encountered in service.

5.1. HAZ Microstructures in $2\frac{1}{4}$Cr–1Mo Steel

The joining of steel components by welding involves remelting of some of the parent material and filling with additional compatible weld metal where necessary. The type, size and shape of the components will influence the choice of welding process, the detailed design of the weld geometry and the manner in which welding is carried out. The normal welding procedures as well as the microstructures of the weld metal and heat affected zones produced during welding of austenitic and ferritic steels have been reviewed previously.[24] For the present purposes, it is therefore sufficient to summarise only the special features associated with the heat affected zones developed during multipass welding of $2\frac{1}{4}$Cr–1Mo steel, a material widely used in the construction of power station and petrochemical plant.

During welding, regions of the steel component immediately adjacent to the fusion boundary are heated rapidly to temperatures which approach the melting temperature of the weld metal and then cooled by conduction at rates determined by heat flow considerations. In these regions, the peak temperatures exceed the Ac_3 temperature of low-alloy ferritic steels so that austenite becomes the stable phase. The subsequent cooling rates are relatively rapid, particularly with thick-section welds in which cooling is essentially complete in around 30 s. The microstructures developed at any position in the heat affected zone of a low-alloy ferritic steel therefore depend on such factors as the exact thermal cycle experienced during the welding operation, the kinetics of austenite formation, and the relevant continuous-cooling-transformation (CCT) behaviour. Since the peak temperatures attained and the subsequent cooling rates decrease with increasing distance from the fusion interface, welding results in a variety of non-equilibrium microstructures in the HAZ of low-alloy ferritic steels.

In the case of $2\frac{1}{4}$Cr–1Mo steel, the normalised and tempered condition is characterised by a mixed ferrite/bainite microstructure with the proportion of bainite varying typically from 30 to 100%. The type and form of the carbides produced by the normalising and tempering operations have been documented by Baker and Nutting.[25] When a single weld bead is deposited on $2\frac{1}{4}$Cr–1Mo steel, the resulting heat affected zone can be considered in relation to three principal regions:

(a) Material near to the fusion boundary attains temperatures well above ~1250 K, dissolving the carbide dispersion which constitutes the main obstacle to growth of the austenite grains. The resulting austenite grain size then depends on the exact time/temperature cycle experienced. On subsequent cooling, the coarse-grained austenite transforms to bainite at ~720 K giving a Widmanstätten bainite lath structure with a coarse-grain size of typically around 100 μm.

(b) At greater distances from the fusion interface, temperatures only just above the Ac_3 are encountered. Although an austenitic grain structure develops, these temperatures are not sufficient to fully dissolve the carbide dispersion inhibiting grain growth. Since transformation initiates more easily with fine- rather than coarse-grained austenite, the grain size is refined by the high rate at which transformation products nucleate on cooling, leading to the formation of a fine-grained bainitic region (grain size ~10 μm).

(c) When the peak temperature falls into the range between Ac_3 and Ac_1, only partial transformation to austenite can occur on heating. Consequently, on cooling, austenite transformation products are found decorating the boundaries surrounding volumes of untransformed material which has merely been tempered during the heating cycle. This is referred to as the 'intercritical' region.

The microstructures developed in the HAZ associated with a single weld

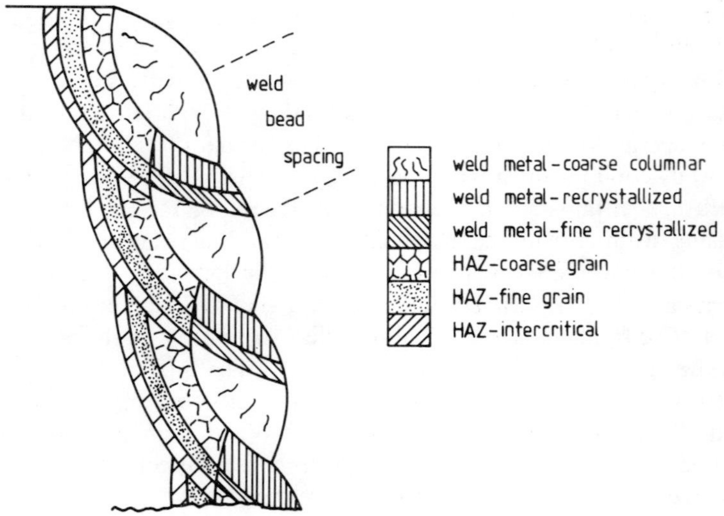

FIG. 10. Schematic illustration of the distribution of structures resulting from the deposition of successive weld beads.[24]

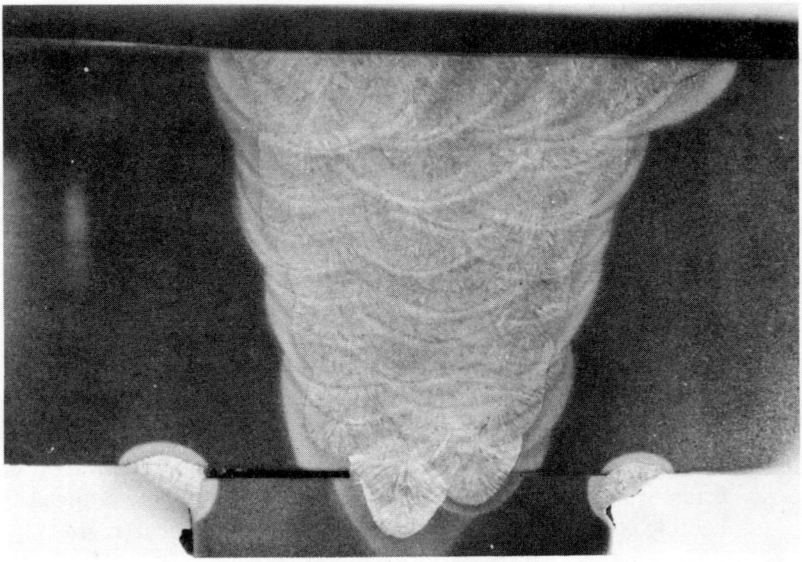

FIG. 11. Microstructure of multipass weld (approximately actual size).

bead laid down on $2\frac{1}{4}$Cr–1Mo steel therefore change predictably with distance from the fusion interface, in a sequence of coarse-grained bainite merging into fine-grained bainite, followed by an intercritically transformed region, and finally tempered parent material. However, with a multipass weld, each successive weld bead heat-treats the underlying material resulting in a complex heat affected zone structure of the form illustrated in Figs 10 and 11.[24]

5.2. Creep Behaviour of Cross-weld Testpieces

Even in the normalised and tempered condition, the microstructures present in $2\frac{1}{4}$Cr–1Mo steel are non-uniform since the size and spacing of the carbides present can vary considerably over distances of the order of the grain size. In the case of $2\frac{1}{4}$Cr–1Mo steel weldments, considerably greater variations in microstructure are easily discernible in the HAZ regions extending several millimetres from the fusion boundary. Moreover, the carbide dispersions initially present in the normalised and tempered parent material and in the various HAZ regions will gradually change with time during creep exposure at elevated temperatures. To add to the problems associated with the complexity of the initial microstructures and the inevitable time-dependent changes in carbide dispersion, the composition of the $2\frac{1}{4}$Cr–1Mo steel in regions adjacent to the fusion boundary may change gradually during high temperature exposure when the composition of the weld metal differs from that of the parent material.

Since the microstructure varies markedly across the heat affected zone formed during multipass welding of $2\frac{1}{4}$Cr–1Mo steel, it must be expected that the creep strength will vary in a corresponding manner. The high temperature performance of a weldment can then be studied by:

(a) Determining the creep properties of the weld metal, the parent steel and a range of parent steel samples subjected to heat treatments designed to simulate each principal type of heat affected zone structure.

(b) Or by comparing the creep curves obtained for the weld metal and parent steel with those for cross-weld testpieces, i.e. for specimens machined from the weldment such that the fusion boundary is located within the gauge length approximately at right angles to the tensile stress axis.

In the present analysis, consideration will be given to both procedures by reference to results obtained for $2\frac{1}{4}$Cr–1Mo steel weldments produced using an austenitic steel filler metal (Armex GT), i.e. for a situation in which the

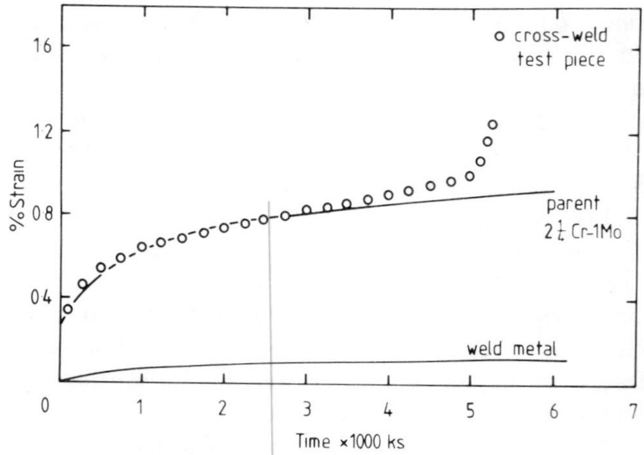

FIG. 12. Creep curve for a cross-weld testpiece compared with that for the normalised and tempered $2\frac{1}{4}$Cr–1Mo steels and the Armex GT weld metal. All tests were performed at 100 MN m^{-2} and 838 K using constant-stress creep equipment.[26]

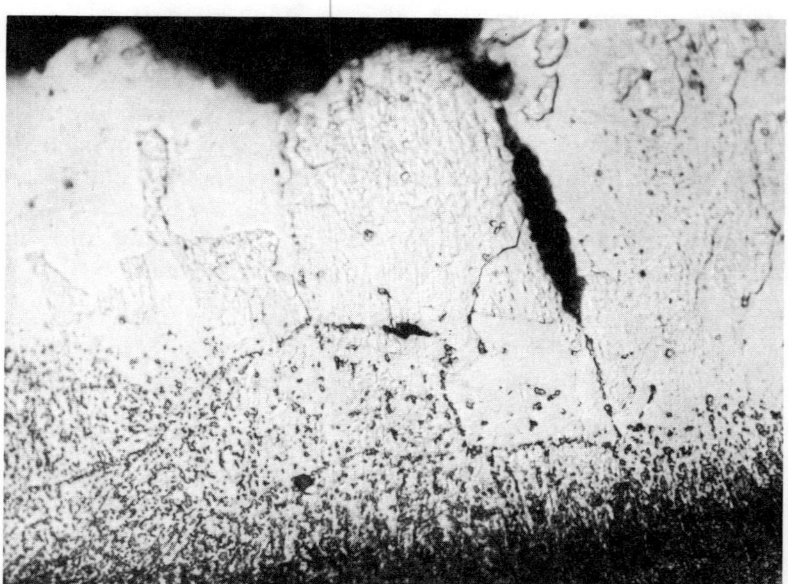

FIG. 13. Micrograph showing the occurrence of fracture at a distance of ~150 μm from the weld interface after creep of a cross-weld testpiece machined from a $2\frac{1}{4}$Cr–1Mo/Armex GT weldment (×500).[26]

high temperature strength of the weld metal is markedly greater than that of the parent steel.[26]

The creep curve recorded for the cross-weld specimen at 100 MN m^{-2} is shown in Fig. 12, together with a portion of the corresponding curves for the $2\frac{1}{4}$Cr–1 Mo steel in the normalised and tempered condition, and for the Armex GT weld metal. Although the cross-weld sample failed in little more than 5×10^3 ks, the tests carried out with the parent $2\frac{1}{4}$Cr–1 Mo steel and the Armex GT weld metal were discontinued when it was established that the creep rate recorded for these materials was still decaying even after periods twice as long as the rupture life of the cross-weld testpiece under these conditions. Microstructural examination of the failed cross-weld sample then confirmed that fracture had occurred in an intergranular manner as a result of crack formation in the coarse-grained region of the heat affected zone of the $2\frac{1}{4}$Cr–1 Mo steel, almost immediately adjacent to the fusion boundary (Fig. 13).

In contrast to the premature low-ductility failures observed for the cross-weld specimens at low stress levels, the creep lives and the rupture ductilities of the cross-weld testpieces were similar to those recorded for the normalised and tempered $2\frac{1}{4}$Cr–1 Mo steels in tests carried out at 200 MN m^{-2} and above. In order to investigate this dramatic difference in the creep fracture behaviour, a study was made of the variation in microstrain distribution along the gauge length of the cross-weld samples tested at high and low stresses. This was accomplished by measuring, after a known creep strain, the detailed displacements apparent on an initially regular surface grid, i.e. the local strains were determined as a function of distance from the fusion boundary by measuring the separations of ~8 μm alumina squares initially ~3·5 μm apart.[27] As shown in Fig. 14, brittle intergranular fractures occur at low stresses when the creep strain accumulates predominantly in regions of the heat affected zone close to the fusion boundary.

The reason for this strain localisation is apparent from the results displayed in Fig. 15. Over the stress range studied for the normalised and tempered $2\frac{1}{4}$Cr–1 Mo steel, the stress dependence of the minimum creep rate can be approximated by a power law as

$$\dot{\varepsilon}_{m} \propto \sigma^{n}$$

with the stress exponent, n, decreasing from ~12 at high stresses to ~4 at stresses below 140 MN m^{-2}. However, no corresponding change in stress exponent was found for the simulated HAZ structures. As a result, stress/creep-rate lines for the parent material and the HAZ structures

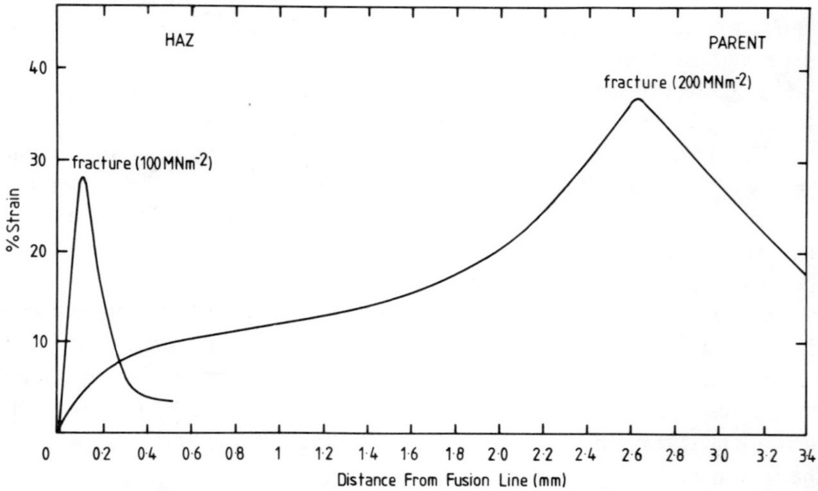

FIG. 14. Distribution of creep strain as a function of distance from the fusion interface during creep of cross-weld testpieces machined from a $2\frac{1}{4}$Cr–1 Mo/Armex GT weldment for tests performed at 100 and 200 MN m^{-2} at 873 K.[26]

intersect at a stress of about 150 MN m^{-2} at 873 K. Thus a marked localisation of creep strain near the fusion boundary (Fig. 14) and brittle intergranular crack formation at a distance of only ~100 to ~150 μm from the weld interface (Fig. 13) are observed for stress levels at which the creep resistance of the HAZ regions is significantly inferior to that of the parent $2\frac{1}{4}$Cr–1 Mo steel.

5.3. Transition Joint Failure—A Case Study

In power generating plant, expensive high-alloy austenitic stainless steels are used in the superheater and reheater regions of steam boilers which operate at high temperatures in a corrosive environment. For economic reasons, ferritic steels are employed where temperatures are lower and this necessitates the use of austenitic/ferritic dissimilar metal welds. Since the mid 1970's there has been worldwide recognition that transition joints can fail well before the design life of the boiler is achieved, i.e. failures become increasingly prevalent after about 15 000 h of service.[26]

Transition joints have a characteristic circumferential failure mode, cracking being located on the ferritic side of the weld immediately adjacent to the fusion boundary, but the detailed mode of fracture depends on whether austenitic or nickel-base filler materials are employed. With austenitic weld metal, damage is accumulated in a narrow band of the heat

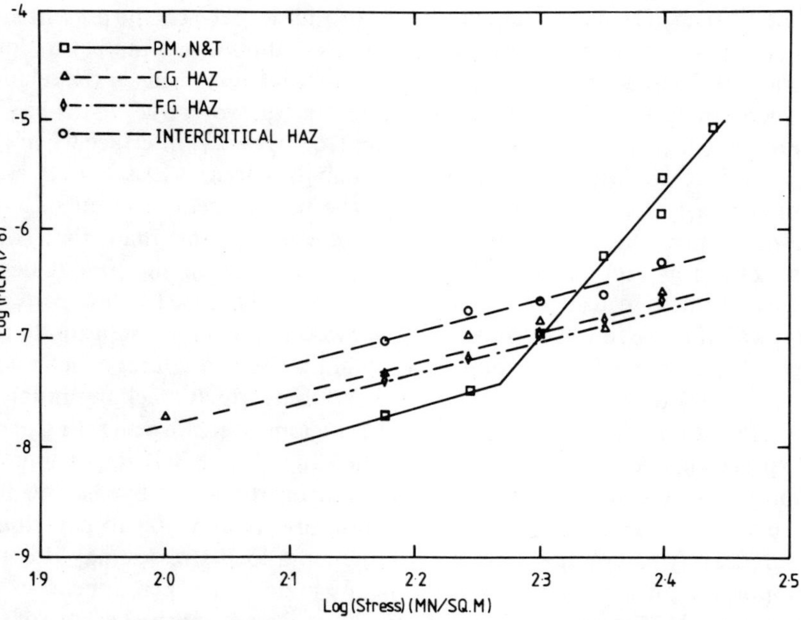

FIG. 15. Variation of the minimum creep rate with stress at 838 K for $2\frac{1}{4}$Cr–1Mo steel in the normalised and tempered condition and for various heat affected zone structures produced by simulative heat treatments.[26]

affected zone of the low-alloy ferritic steel giving rise to intergranular crack formation at a distance of approximately 100 μm from the fusion line. In contrast, examination of failures at welds between type 321 austenitic steel and $2\frac{1}{4}$Cr–1Mo ferritic steel produced using Inconel 132 filler metal reveals a different mode of crack development. With the nickel-base weld metal, cracks develop by link-up of voids which originate at coarse carbide precipitates formed in the ferritic steel at a distance typically some 5 μm from the fusion line.[26]

Although considerable attention has been directed towards the classification of the parameters characterising transition joint failure in service, the detailed processes by which cracking develops have remained uncertain. Since the exact failure mode depends on whether austenitic or nickel-base weld metals are used, no one mechanism of fracture will be applicable to all types of transition joint. Even so, the fact that the characteristics associated with failure of transition joints produced using austenitic weld metal can be reproduced in cross-weld specimens by performing uniaxial creep tests at low stresses does illustrate the importance of creep fracture processes. In addition, the results presented in Figs 14 and 15 show that a qualitative explanation for the premature low-ductility failure of transition joints can be provided in terms of the relative creep strengths of the parent steel and the various HAZ structures.[26] Failure occurs in regions only $\sim 100\ \mu m$ from the weld interface when the HAZ structures are less creep-resistant than the parent steel at low stresses. In contrast, ductile fracture occurs in the parent metal well outside the HAZ when the parent steel is less creep-resistant than the HAZ structures at high stresses. Thus attempts to shorten the time-scale of the creep tests by using higher applied stresses fail to give fracture patterns typical of those found during service. However, a complete solution to the problem requires the development of quantitative procedures which allow predictions to be made of the exact location at which cracking initiates.

Transition joints are usually located in regions of steam generating plant experiencing service temperatures in the range 770–870 K depending on their exact location. The stresses imposed on the joints by the 160 bar ($16\ MN\ m^{-2}$) steam pressure are low and are thought not to contribute significantly to eventual failure. Although end load stresses may become important during service, the stresses of major relevance appear to be generated by thermal transients. These thermal stresses arise because of the mismatch in thermal expansion coefficients between the austenitic and ferritic materials. Obviously, the exact conditions encountered during service of transition joints are not known nor are the relevant creep properties as a function of temperature. Consequently, at this juncture, predictions cannot be made for the behaviour patterns expected for transition joints during plant operation. Even so, it is possible to see whether numerical procedures can be developed to predict the strain inhomogeneities (Fig. 14) and fracture mode observed (Fig. 13) for cross-weld testpieces tested under conditions which simulate transition joint failure in service. (Limitation of the numerical studies to the situation of cross-weld testpieces which fail in times which are short compared with service life also has the advantage of eliminating the need to make allowance in terms of creep properties for the time-dependent changes in composition which occur near the fusion interface as diffusion of alloying elements takes place between the austenitic and ferritic steels.)

Figure 16(a) shows a finite element mesh idealisation of a cross-weld testpiece. The model is for the axisymmetric condition and uses eight-noded isoparametric quadratic elements with reduced 2×2 integration to maintain incompressibility (see Section 4.3). The materials are assumed to obey Norton's law with the values for the various constants A and n taken

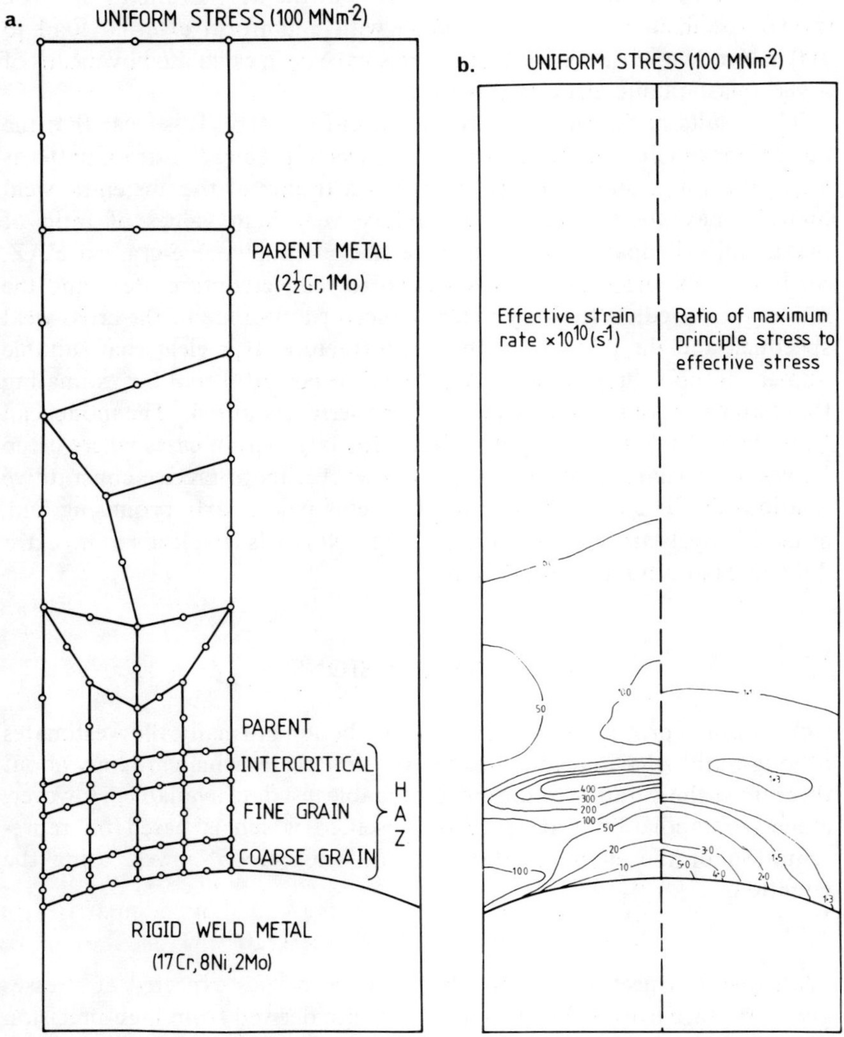

FIG. 16. Finite element modelling for the creep of a cross-weld testpiece machined from a $2\frac{1}{4}$Cr–1Mo/Armex GT weldment at 838 K. (a) The finite element mesh. (b) Contours of effective strain rate and of the ratio of maximum principal stress to effective stress.

from Fig. 15. The series of fine-scale elements near to the rigid austenitic weld metal model the various components of the heat affected zone. The model specimen is loaded in tension with a uniform external load to 100 MN m^{-2} and the various creep rates have been calculated by means of a viscoplastic finite element program.

The results of this analysis are shown in Fig. 16(b). It is clear that the conditions of effective strain rate are such as to predict the strain patterns along the gauge length. However, the constraints of the austenitic weld material have been such as to produce very high values of ratio of maximum principal stress to effective stress in the coarse-grained HAZ. Such a combination of stress is known to reduce rupture life[23] and the model then predicts the very much reduced rupture life of the cross-weld specimen and the position of the creep fracture. It is clear that suitable numerical modelling at the design stage is a powerful tool for estimating those areas of components which may be seriously at risk. The model can be improved substantially, particularly for large strain cases where creep curves are tertiary dominated, by the use of a more precise constitutive relationship. The θ function (eqn (7)) seems particularly promising and, although algebraically more complex than Norton's law, leads to no extra difficulties in a numerical analysis.

6. CONCLUSIONS

Conventional procedures for creep life prediction normally allow estimates of long-term behaviour to be made over time-scales which are only about three times the duration of the longest reliable test data available. However, a new extrapolation procedure is presented which is based on representation of the shape of the constant-stress creep curves, using the equation

$$\varepsilon = \theta_1[1 - \exp(-\theta_2 t)] + \theta_3[\exp(\theta_4 t) - 1]$$

With this θ projection concept the creep properties expected at stresses giving rupture lives of 10 years and more are derived from high-precision data obtained in less than three months. In addition to predicting stress-rupture behaviour, this new extrapolation procedure also offers full characterisation of the creep properties of engineering materials so that modern design methods can be adopted for high temperature design.

If the predictions of the θ projection concept can be validated by laboratories concerned with the acquisition of long-term data, it may prove possible to extend this procedure to provide a new method of remanent life assessment and even the data range required in order that the behaviour of welded structures can be modelled using finite element methods. Indeed, even using a more limited range of materials data, the precise microstrain distributions and the exact fracture modes observed during creep of cross-weld samples taken from austenitic/ferritic steel transition welds can be forecast using viscoplastic finite element programs.

REFERENCES

1. Evans, R. W., Parker, J. D. and Wilshire, B. (1982). In: *Recent Advances in Creep and Fracture of Engineering Materials and Structures* (eds B. Wilshire and D. R. J. Owen), Pineridge Press, Swansea, p. 135.
2. Goodman, A. M. (1981). In: *Creep and Fatigue in High Temperature Alloys* (ed. J. Bressers), Applied Science Publishers Ltd, London, p. 145.
3. Browne, R. J., Cane, B. J., Parker, J. D. and Walters, D. J. (1981). In: *Creep and Fracture of Engineering Materials and Structures* (eds B. Wilshire and D. R. J. Owen), Pineridge Press, Swansea, p. 645.
4. Johnson, R. F. and Glen, J. (1974). *Creep Strength of Steels and High Temperature Alloys*, Metals Society, London, p. 37.
5. Cane, B. J. (1982). *Int. J. Press. Vess. Piping*, **10**, 11.
6. Ashby, M. F. (1972). *Acta Met.*, **20**, 887.
7. Taylor, R. R. and Johnson, R. F. (1971). *JISI*, **209**, 714.
8. Mendelson, A., Roberts, E. and Manson, S. S. (1965). NASA-TN-D-2975.
9. Manson, S. S. and Haferd, A. V. (1953). NACA-TN-2890.
10. Larson, F. R. and Miller, J. (1952). *Trans ASME*, **74**, 765.
11. Goodman, A. M., Orr, J. and Draper, J. H. M. (1980). *Engineering Aspects of Creep*, Inst. Mech. Eng., London, p. 159.
12. Etienne, C. F. and van Helst, H. C. (1978). In: *Creep of Engineering Materials and Structures* (eds G. Bernasconi and G. Piatti), Applied Science Publishers Ltd, London, p. 149.
13. Robinson, E. L. (1938). *Trans. ASME*, **60**. 253.
14. Birch, J. M., Wilshire, B., Owen, D. R. J. and Shantaram, D. (1976). *J. Mater. Sci.*, **11**, 1817.
15. Williams, K. R. (1981). In: *Creep and Fracture of Engineering Materials and Structures* (eds B. Wilshire and D. R. J. Owen), Pineridge Press, Swansea, p. 489.
16. Dyson, B. F. and McLean, D. (1972). *Metal Sci. J.*, **6**, 220.
17. Auerkari, P. (1983). In: *Advances in Life Prediction Methods* (eds D. A. Woodford and J. R. Whitehead), ASME, New York, p. 353.
18. Neubauer, B. (1981). In: *Creep and Fracture of Engineering Materials and Structures* (eds B. Wilshire and D. R. J. Owen), Pineridge Press, Swansea, p. 617.
19. Plastow, B. and Davison, J. To be published.

20. Williams, K. R. and Wilshire, B. (1981). *Mater. Sci. Eng.*, **47**, 151.
21. Graham, A. and Walles, K. F. A. (1955). *JISI*, **193**, 105.
22. Hart, E. W. (1975). *Constitutive Equations in Plasticity* (ed. A. S. Argon), MIT Press, Cambridge, p. 149.
23. McLean, D., Dyson, B. F. and Taplin, D. M. R. (1977). In: *Fourth International Conference on Fracture* (ed. D. M. R. Taplin), Waterloo Press, Waterloo, p. 325.
24. Price, A. T. and Williams, J. A. (1982). In: *Recent Advances in Creep and Fracture of Engineering Materials and Structures* (eds B. Wilshire and D. R. J. Owen), Pineridge Press, Swansea, p. 265.
25. Baker, R. G. and Nutting, J. (1959). ISI Spec. Rep. 64, p. 1.
26. Chilton, I. J., Price, A. T. and Wilshire, B. (1984). *Metals Technology*, in press.
27. Parker, J. D. and Wilshire, B. (1977). *Mater. Sci. Eng.*, **29**, 219.

5

On the Creep and Viscoplastic Lifetime Prediction of Structures

D. R. J. OWEN
*Department of Civil Engineering,
University College of Swansea, UK*

and

O. J. A. GONÇALVES F$^{O.}$
*Instituto de Engenharia Nuclear—CNEN,
Rio de Janeiro, Brazil*

1. INTRODUCTION

The simultaneous demands for higher pressures and temperatures in order to improve the thermodynamic efficiency of power generating plants have led to an increased need for accurate methods of stress prediction in complex structures under transient loading conditions. Numerical techniques, in particular the finite element method, have proved to be invaluable in this respect.

The problems of life prediction encountered in power plant design almost invariably require that the effects of plastic deformation be included in the structural analysis. The plastic deformation of materials is basically a time-rate-dependent phenomenon. In the classical theory of plasticity,[1] such rate effects are ignored and the material performance is described incrementally under assumed conditions of instantaneous quasi-static plastification. The theory of viscoplasticity[2-5] attempts to provide a material model more in keeping with experimental observations by introducing rate dependence into the constitutive relations. Whether or not the inclusion of such effects has a significant influence on the prediction of the material behaviour depends upon several factors. In the study of structural components under static loading at normal temperatures it is accepted that time-rate effects are generally not important and the conventional theory of plasticity then models the behaviour adequately. However, many structural materials, especially under high operating temperatures, exhibit a phenomenon of stress and/or strain redistribution with time which may be either elastic (creep) or plastic in nature. The theory of elasto-viscoplasticity permits the numerical analysis of such problems; quite complex material responses can be simulated with the aid of 'overlay models', as described in Ref. 6.

The engineering approaches formerly adopted to estimate the rupture time of structures undergoing creep deformation were based on determining the elastic or the steady-state stress distribution and using the corresponding maximum principal stress component together with the material uniaxial rupture data, such as those given in ASME Code Case N47-12. These approaches have proved deficient on two accounts by ignoring, firstly, the influence of multiaxial stress states upon the rupture time and, secondly, the softening effect exhibited by metals over the tertiary creep range which allows stresses to be transferred from a highly damaged to a less damaged zone which, consequently, increases the service life of the structure.

In order to overcome these deficiencies, Kachanov[7] and Rabotnov[8] introduced the concept of material damage in the constitutive equations in the form of a scalar state variable referred to as the damage parameter w. Since the original propositions by Kachanov and Rabotnov for creep rupture, several studies conducted in different laboratories have shown the extensive possibilities of their original approach. These studies initiated the development of a new branch of fracture mechanics termed continuous damage mechanics (CDM).

Material damage and fracture are generally induced by nucleation of microcavities and their growth and coalescence into microscopic cracks in the material. The effects of these internal material defects may be reasonably discussed by dividing them into a single or finite number of macroscopic cracks, and a set of distributed microcavities preceding these macroscopic cracks. This is the approach followed in classical fracture mechanics.

The presence of distributed defects in the material body, on the other hand, not only leads to crack initiation and final fracture, but also induces material deterioration such as decrease of strength, rigidity, toughness and stability, and, ultimately, a decrease of residual life. Investigation of the growth of microcavities and the mechanical behaviour of damaged materials are the objectives of continuous damage mechanics.[9,10]

The Kachanov–Rabotnov theory mentioned above has provided an important basis for CDM and has served as a prototype for many of the damage theories developed thereafter. Leckie and Hayhurst,[11] for instance, generalised the one-dimensional Kachanov–Rabotnov equations to multiaxial stress states by assuming isotropy of the material and of the damage process. Because of these assumptions their generalisation should only be applied to proportionally loaded structures.

Recent experimental investigations indicate that for complex stress histories due to non-proportional loading, voids and cracks on the grain boundaries develop predominantly on planes perpendicular to the maximum principal stress. Clearly for these cases more elaborate theories are required introducing a damage tensor instead of a scalar.

The development of a consistent theory for anisotropic damage poses some difficulties, especially in connection with thermodynamic requirements.[12] Hitherto some theories describing material damage in terms of tensorial variables have been proposed for the study of creep and viscoplastic rupture.[12–16] Two of the most attractive theories are due to Murakami and Ohno,[13] who introduced damage as a symmetric second-rank tensor, and to Chaboche,[14] who defined damage as an asymmetrical fourth-rank tensor.

So far, few experiments have been conducted to try to identify the relationships between the tensorial variables introduced and the actual mechanisms of microstructural change in the materials. One of the most interesting results arrived at, is that the effect of material damage on creep deformation may be assumed to be isotropic when the cavity density is less than a certain amount.[17] This result is of particular importance because it allows a greater simplicity in the constitutive equations.

Considering that the experiments required to validate the tensorial theories are both time-consuming and expensive, Hayhurst *et al.*[18] have developed an experimental programme to assess the validity of using the isotropic theory for non-proportional loading conditions. The experimental results showed that for engineering cases of non-proportional loading the creep deformation and rupture behaviour of the materials studied can be described by the scalar variable theory, provided that the directional characteristics of the damage process are taken into account during the integration of the damage rate equations.

In this paper, the basic laws governing time-dependent plastic continuum behaviour are first summarised before considering numerical formulation of the problem. Elasto-creep damage behaviour is modelled using the multiaxial isotropic theory of Leckie and Hayhurst while elasto-viscoplastic damage behaviour is described by a simplified isotropic theory introduced by the authors.[19]

An implicit algorithm (Runge–Kutta scheme of second-order) is then developed for the finite element analysis of creep and viscoplastic brittle rupture. Geometric non-linear effects, which may play an important role in design against rupture, are accounted for by means of an updated Lagrangian formulation. Special attention is given to the solution of the highly non-linear differential equations that govern damage growth. The computational scheme developed is illustrated by the investigation of the creep damage and rupture of a spherical pressure vessel with a cylindrical nozzle junction submitted to a constant load histogram.

2. THEORY OF ELASTO-VISCOPLASTICITY WITH INTERNAL STATE VARIABLES

The non-linear behaviour of real materials originates from both plastic and creep internal dissipation of energy and it is difficult to distinguish between the two effects. Some of the earlier theories which attempted to simultaneously consider plasticity and time effects can be found in Refs 20–22. On this basis, Perzyna and Olszak[2–5] proposed the rate-dependent elasto-viscoplastic model as a more realistic simulation of the inelastic behaviour of a large spectrum of materials.

In this model the total strain tensor ε_{ij}, and consequently the strain-rate tensor, can be decomposed into its elastic and inelastic (viscoplastic) parts according to

$$\varepsilon_{ij} = \varepsilon_{ij}^{\mathrm{e}} + \varepsilon_{ij}^{\mathrm{vp}} + \varepsilon_{ij}^{0} \tag{1}$$

in which $\varepsilon_{ij}^{\mathrm{e}}$, $\varepsilon_{ij}^{\mathrm{vp}}$, ε_{ij}^{0} represent, respectively, the elastic, viscoplastic and initial strain components.

In order to formulate a theory which models elasto-viscoplastic material deformation, three requirements have to be met:

(a) An explicit relationship between stress and strain must be formulated to describe material behaviour under elastic conditions, i.e. before the onset of plastic deformation.
(b) A yield criterion indicating the stress level at which plastic flow commences must be postulated.
(c) A relationship between stress and strain must be developed for post-yield behaviour, i.e. when the deformation is made up of both elastic and viscoplastic components.

The elastic strain is related to the total stress according to the generalised Hooke's law

$$\varepsilon_{ij}^{\mathrm{e}} = \frac{S_{ij}}{2\mu} + \frac{(1-2\nu)}{E}\sigma_{\mathrm{m}}\delta_{ij} \tag{2}$$

in which $S_{ij} = \sigma_{ij} - \delta_{ij}\sigma_m$ is the deviatoric stress tensor, $\sigma_m = 1/3\sigma_{ii}$ is the mean hydrostatic pressure, and μ, E and ν are the shear modulus, elastic modulus and Poisson's ratio, respectively.

Equation (2) completely describes the reversible part of the deformation. The viscoplastic response of the material becomes manifest as soon as some specified combination of the stress components at a point exceeds a characteristic value. This behaviour is governed by a scalar yield function as described in the next section.

2.1. The Yield Criterion

The yield criterion determines the stress level at which plastic deformation begins and can be written in the general form

$$f(\sigma_{ij}) = f(J_2', J_3') = Y(k) \tag{3}$$

where f is some function of the deviatoric stress invariants

$$J_2' = \tfrac{1}{2}S_{ij}S_{ij} \tag{4}$$

$$J_3' = \tfrac{1}{2}S_{ij}S_{jk}S_{ki} \tag{5}$$

The term Y in eqn (3) can be a function of a hardening parameter k. The two most common yield criteria employed in the description of the behaviour of metals are the Tresca criterion and the von Mises criterion.

2.1.1. The Tresca Yield Criterion

This states that yielding begins when the maximum shear stress reaches a certain value. If the principal stresses are $\sigma_1, \sigma_2, \sigma_3$, where $\sigma_1 \geq \sigma_2 \geq \sigma_3$, then yielding begins when

$$\sigma_1 - \sigma_3 = Y(k) \tag{6}$$

where Y is a material parameter to be experimentally determined and which may be a function of the hardening parameter k. By considering all other possible maximum shearing stress values (e.g. $\sigma_2 - \sigma_1$ if $\sigma_2 \geq \sigma_3 \geq \sigma_1$) it can be shown that this yield criterion may be represented in the $\sigma_1\sigma_2\sigma_3$ stress space by the surface of an infinitely long regular hexagonal cylinder. The axis of the cylinder coincides with the space diagonal, defined by points $\sigma_1 = \sigma_2 = \sigma_3$.

2.1.2. The Von Mises Criterion

Von Mises suggested that yielding occurs when J_2' reaches a critical value, or

$$(J_2')^{1/2} = K(k) \tag{7}$$

in which K is a material parameter to be determined. The second deviatoric stress invariant, J_2', can be explicitly written as

$$J_2' = \tfrac{1}{2}S_{ij}S_{ij} = \tfrac{1}{6}[(\sigma_1 - \sigma_2)^2 + (\sigma_2 - \sigma_3)^2 + (\sigma_3 - \sigma_1)^2] \tag{8}$$

Yield criterion (7) may be further written as

$$\bar{\sigma} = \sqrt{(3)}(J_2')^{1/2} = \sqrt{(3)}K \tag{9}$$

where

$$\bar{\sigma} = \sqrt{(\tfrac{3}{2})}\{S_{ij}S_{ij}\}^{1/2} \tag{10}$$

and $\bar{\sigma}$ is termed the effective stress, generalised stress or equivalent stress. The geometrical interpretation of the von Mises yield surface can be shown to be a circular cylinder of radius $\sqrt{(2)}K$ aligned with the space diagonal. A physical meaning of the constant K can be obtained by considering the yielding of materials under simple stress states. The case of pure shear ($\sigma_1 = -\sigma_2$, $\sigma_3 = 0$) requires, on use of eqns (7) and (8), that k must equal the yield shear stress. Alternatively the case of uniaxial tension ($\sigma_2 = \sigma_3 = 0$) requires that $\sqrt{(3)}K$ be the uniaxial yield stress.

The Tresca yield locus is a hexagon with distance of $\sqrt{(\tfrac{2}{3})}Y$ from origin to apex on the π plane whereas the von Mises yield surface is a circle of radius $\sqrt{(2)}K$. By suitably choosing the constant Y, the criteria can be made to agree with each other, and with experiment, for a single state of stress. This may be selected arbitrarily; it is convenient to make the circle pass through the apices of the hexagon by taking the constant $Y = \sqrt{(3)}K$, the yield stress in simple tension. The criteria then differ most for a state of pure shear, where the von Mises criterion gives a yield stress $2/\sqrt{(3)}$ ($\simeq 1{\cdot}15$) times that given by the Tresca criterion.

2.2. Work or Strain Hardening

After initial yielding, the stress level at which further plastic deformation occurs may be dependent on the current degree of plastic straining. Thus the yield surface will vary at each stage of the plastic deformation with the subsequent yield surfaces being dependent on the plastic strains in some way. In this work, attention is restricted to an isotropic hardening model in which the original yield surface expands uniformly without translation.

The progressive development of the yield surface can be defined by relating the yield stress Y to the plastic deformation by means of the hardening parameter k. In a work-hardening hypothesis k is related to the total plastic work, w_p, as

$$k = w_p = \int \sigma_{ij}(\mathrm{d}\varepsilon_{ij})_p \tag{11}$$

in which $(\mathrm{d}\varepsilon_{ij})_\mathrm{p}$ are the plastic components of strain occurring during a strain increment. Alternatively, in a strain-hardening hypothesis, k is related to a measure of the total plastic deformation termed the effective or equivalent plastic strain which is defined incrementally as

$$\mathrm{d}\bar{\varepsilon}_\mathrm{p} = \sqrt{(\tfrac{2}{3})}\{(\mathrm{d}\varepsilon_{ij})_\mathrm{p}(\mathrm{d}\varepsilon_{ij})_\mathrm{p}\}^{1/2} \tag{12}$$

Then

$$k = \bar{\varepsilon}_\mathrm{p} \tag{13}$$

where $\bar{\varepsilon}_\mathrm{p}$ is the result of integrating $\mathrm{d}\bar{\varepsilon}_\mathrm{p}$ over the strain path.

The yield criteria given by eqns (6) and (7) can then be expressed in the general form

$$F = f(\sigma_{ij}, k) - Y(k) = 0 \tag{14}$$

in which Y is the uniaxial or effective yield stress and values of $F < 0$ imply an elastic state.

2.3. Elasto-viscoplastic Stress–Strain Relationships

A very general expression for the viscoplastic strain rate under isothermal conditions is given by

$$\dot{\varepsilon}_{ij}^{\mathrm{vp}} = \dot{\varepsilon}_{ij}^{\mathrm{vp}}(\sigma_{ij}, \alpha_n) \tag{15}$$

where σ_{ij} are the Cauchy stresses, α_n is a set of n variables (scalers or tensors) describing the structural state of the material and the superposed dot indicates time differentiation.

In this work, two internal state variables are considered, namely strain hardening (k) and damage (w). The first variable has already been introduced while the second one will be discussed in detail in the next section.

One specific form of eqn (15) which has wide applicability is given by the following flow rule

$$\dot{\varepsilon}_{ij}^{\mathrm{vp}} = \gamma_\mathrm{f}\langle\Phi(f, Y, w)\rangle\frac{\partial F}{\partial\sigma_{ij}} \tag{16}$$

where f is the von Mises or the Tresca stress, Y is the current value of the yield stress, and F ($F = f - Y$) is a potential function. The notation $\langle\Phi\rangle$ implies that $\langle\Phi\rangle = 0$ if $\Phi \leq 0$.

Finally the total rate of stress change in an elasto-viscoplastic medium can be expressed as

$$\dot{\sigma}_{ij} = C_{ijkl}\dot{\varepsilon}_{kl}^{\mathrm{e}} = C_{ijkl}(\dot{\varepsilon}_{kl} - \dot{\varepsilon}_{kl}^{\mathrm{vp}}) \tag{17}$$

where C_{ijkl} is the usual constitutive tensor for elastic materials.

2.4. Damage Evolution Equations and Rupture Criteria

Based on the experimental results of Johnson *et al.*[23] for tension–torsion tests performed on copper and aluminium tubes, Leckie and Hayhurst[11] proposed the following equations to describe the creep strain rate and the damage rate, respectively

$$\dot{\varepsilon}_{ij}^{c} = \gamma_f \left(\frac{| \bar{\sigma} |}{1 - w} \right)^N \frac{\partial F}{\partial \sigma_{ij}} \tag{18}$$

and

$$\dot{w} = A \frac{\sigma_{eq}^{\chi}}{(1 - w)^{\phi}} \tag{19}$$

where ε_{ij}^{c} is the tensor of infinitesimal creep strains, σ_{ij} is the Cauchy stress tensor, $\bar{\sigma}$ is the von Mises or Tresca stress, F is a potential function, γ_f, N, χ, A and ϕ are temperature-dependent properties, and σ_{eq} is a scalar quantity that takes account of the rupture characteristics of the material under multiaxial stress states. In a general form

$$\sigma_{eq} = \alpha_1 \sigma_1 + \alpha_2 \tau_{max} + \alpha_3 J_1 + \alpha_4 \sigma_{eff} \tag{20}$$

where σ_1 is the maximum principal tensile stress, τ_{max} is the maximum shear stress, J_1 is the first stress invariant, σ_{eff} is the effective stress, and α_i $(i = 1, 4)$ are scalar quantities such that $(\alpha_1 + \alpha_2 + \alpha_3 + \alpha_4 = 1{\cdot}0)$. The cases $(\alpha_1 = 1,\ \alpha_2 = \alpha_3 = \alpha_4 = 0)$ and $(\alpha_1 = \alpha_2 = \alpha_3 = 0,\ \alpha_4 = 1)$ represent the extreme types of rupture behaviour; materials which fail under these conditions are said to obey a maximum principal tension stress rupture criterion and a maximum effective tension stress criterion, respectively. Copper $(\alpha_1 = 1)$ and aluminium alloys $(\alpha_4 = 1)$ are typical examples of extreme behaviour while the behaviour of other alloys, particularly creep-resistant ones, falls between the two extremes. In the initial undamaged state $w = 0$; at rupture $w = 1$.

In the case of elasto-viscoplastic strain-hardening materials a simplified procedure for including damage in the constitutive equations has been suggested.[19] Damage is introduced as a scalar variable and a direct coupling between hardening and damage (through their respective evolutional equations) is considered unimportant.

The constitutive and evolutional equations proposed in Ref. 19 to describe elasto-viscoplastic deformation in multiaxial stress states are

$$\dot{\varepsilon}_{ij}^{vp} = \gamma_f \left\langle \left(\frac{f - Y}{(1 - w)Y} \right)^N \right\rangle \frac{\partial F}{\partial \sigma_{ij}} \tag{21}$$

$$\dot{w} = A \left\langle \frac{\hat{\sigma}_{eq}(f, Y)^{\chi}}{(1 - w)^{\phi}} \right\rangle \tag{22}$$

and

$$Y = \sigma_y + H'\bar{\varepsilon}^{vp} \tag{23}$$

where σ_y is the material uniaxial yield stress, H' is the slope of the uniaxial stress versus plastic strain curve, and the remaining quantities have been previously defined. The circumflex indicates that σ_{eq} is calculated with reference to the resulting overstress only. The tensor of overstress acting in any particular point subject to a stress state σ_{ij} is given by

$$\hat{\sigma}_{ij} = \left\langle 1 - \frac{Y}{f} \right\rangle \sigma_{ij} \tag{24}$$

As discussed before, damage is assumed to be the predominant phenomenon during the tertiary stage of the creep/viscoplastic deformation process (strain-hardening recovery is ignored) and a direct coupling between damage and hardening is considered unimportant. To account for these factors in the constitutive equations (21) to (23) the expansion of the yield surface Y is stopped at any point in the solid where the damage state begins to be meaningful ($w \geq 0{\cdot}01$).

In concluding this section it is worth noting that the constitutive equations (18) to (20) proposed by Leckie and Hayhurst[11] for the analysis of creep damage and rupture in multiaxial stress states are readily obtained as a particular case of the more general elasto-viscoplastic model (eqns (21)–(23)) by setting the threshold yield stress σ_y and the hardening parameter H' to zero.

3. FINITE ELEMENT APPROXIMATION

In this section, the finite element method[24,25] is applied to develop the numerical solution of the quasi-static elasto-viscoplastic problem of interest. The principle of virtual work is invoked to express the overall equilibrium condition at any instant and the finite element equations are derived according to a displacement formulation. As observed in Section 1, geometric non-linear effects (mainly large displacements and large rotations) may assume, at least for select applications, an important role in the design of metallic components against brittle failure[26] and should not be disregarded in rupture calculations. These effects are also included in this work as discussed in the next section.

3.1. Problem Formulation

First it is noted that the governing differential equations must always refer to the current structure configuration at time t, Ω_t, which is assumed to be a bounded region in R^n ($n = 2, 3$) with a piecewise smooth boundary Γ_t. As this configuration is not known *a priori*, the solution is obtained by using an incremental procedure. In principle, any of all the previous configurations could be used as a reference configuration but, in practice, the choice lies essentially between two approaches which are termed total Lagrangian (TL) and updated Lagrangian (UL) formulations.[27] In the TL formulation all variables are referred to the initial undeformed configuration while in the UL description all variables are referred to the last calculated configuration. Clearly, if appropriate transformations are performed both formulations yield the same numerical results. Therefore the choice of a formulation depends basically on its numerical efficiency for the solution of the problem under consideration.

In the present study the UL formulation has been adopted. This formulation leads to simpler strain–displacement expressions and is probably the most natural to use. As is common in most available computer programs, the material constants are not transformed when updating the reference configuration. These transformations increase the computer time drastically and, moreover, as they must be performed at each integrating point they induce an artificial anisotropy into the material. The omission of the material constant transformations is acceptable provided that the strains remain small throughout the response.

The problem is now formulated in an incremental form. First let the solution be known at all discrete time stations 0, $t_1, t_2, \ldots, t_n$. Then by considering the principle of virtual work the overall equilibrium condition at time t_{n+1} is expressed as

$$\int_{\Omega_{t_{n+1}}} (\sigma_{ij})_{n+1}(\delta\varepsilon_{ij})_{n+1}\,\mathrm{d}\Omega = (\delta W_e)_{n+1} \tag{25}$$

where σ_{ij} is the Cauchy stress tensor, ε_{ij} is the tensor of infinitesimal strains, W_e is the work done by the external forces and the symbol δ denotes a virtual variation. The external virtual work term δW_e is given by

$$(\delta W_e)_{n+1} = \int_{\Omega_{t_{n+1}}} (b_i)_{n+1}(\delta\mu_i)_{n+1}\,\mathrm{d}\Omega + \int_{\Gamma_{2_{t_{n+1}}}} (\bar{p}_i)_{n+1}(\delta\mu_i)_{n+1}\,\mathrm{d}\Gamma \tag{26}$$

where b_i are the body forces, $\bar{p}_i$ are the prescribed surface tractions on Γ_2, and $\delta\mu_i$ are the virtual displacements.

In eqns (25) and (26) all quantities are referred to the domain configuration at time t_{n+1}. By using the notion of stress–strain energy conjugate,[28,29] eqn (25) is transformed to

$$\int_{\Omega_{t_n}} (S_{ij})_{n+1/n} (\delta e_{ij})_{n+1/n} \, \delta\Omega = (\delta W_e)_{n+1} \tag{27}$$

where S_{ij} is the second Piola–Kirchhoff stress tensor, and e_{ij} is the Green–Lagrange strain tensor. The subscript $n+1/n$ indicates that a quantity at time t_{n+1} is calculated with reference to the last known configuration Ω_{t_n} while the subscript n (or $n+1$) indicates that a quantity is evaluated at time t_n (or t_{n+1}) with reference to the current configuration Ω_{t_n} (or $\Omega_{t_{n+1}}$).

In the UL formulation the Green–Lagrange strains are given by

$$e_{ij} = \varepsilon_{ij} + \eta_{ij} = \tfrac{1}{2}(\mu_{i,j} + \mu_{j,i}) + \tfrac{1}{2}(\mu_{k,i}\mu_{k,j}) \tag{28}$$

and the external virtual work is approximated by

$$(\delta W_e)_{n+1} \doteq \int_{\Omega_{t_n}} (b_i)_{n+1/n} (\delta\mu_i)_{n+1/n} \, d\Omega + \int_{\Gamma_{2_{t_n}}} (\bar{p}_i)_{n+1/n} (\delta\mu_i)_{n+1/n} \, d\Gamma \tag{29}$$

The second Piola–Kirchhoff stresses $(S_{ij})_{n+1/n}$ are written in incremental form as

$$(S_{ij})_{n+1/n} = (\sigma_{ij})_n + (\Delta S_{ij})_{n+1/n} \tag{30}$$

where $(\sigma_{ij})_n$ are the Cauchy stresses at time t_n determined from

$$(\sigma_{ij})_n = \frac{\rho_n}{\rho_{n-1}} (x_{i,k})_{n/n-1} (S_{kl})_{n/n-1} (x_{j,l})_{n/n-1} \tag{31}$$

where ρ_n and ρ_{n-1} are the specific mass of the body at times t_n and t_{n-1}, respectively, and $(x_{i,k})_{n/n-1}$ and $(x_{j,l})_{n/n-1}$ are the derivatives of the Cartesian coordinates x_i and x_j at time t_n with respect to the coordinates x_k and x_l at time t_{n-1}.

The constitutive relationship between the stress and strain increments is obtained from eqn (17) as

$$(\Delta S_{ij})_{n+1/n} = C_{ijkl} (\Delta\varepsilon^E_{kl})_{n+1/n} \tag{32}$$

where C_{ijkl} is the elasticity tensor calculated, as discussed before, at the initial undeformed configuration, and $(\Delta\varepsilon^E_{kl})_{n+1/n}$ is the elastic component of the linear part of the Green–Lagrange strain increment $(\Delta e_{kl})_{n+1/n}$.

Substituting eqns (30) and (32) into eqn (27) and employing the assumption that $(\delta e_{ij})_{n+1/n} \doteq (\delta \varepsilon_{ij})_{n+1/n}$ results in

$$\int_{\Omega_{t_n}} C_{ijkl}(\Delta\varepsilon_{kl}^{E})_{n+1/n}(\delta\varepsilon_{ij})_{n+1/n}\,d\Omega + \int_{\Omega_{t_n}} (\sigma_{ij})_n(\delta\eta_{ij})_{n+1/n}\,d\Omega$$

$$= (\delta W_e)_{n+1} - \int_{\Omega_{t_n}} (\sigma_{ij})_n(\delta\varepsilon_{ij})_{n+1/n}\,d\Omega \quad (33)$$

which is, finally, the incremental form sought for the original initial boundary value problem.

3.2. Finite Element Incremental Equations

Using the isoparametric finite element discretisation concept[24,25] to evaluate the displacement derivatives in the integrals, the finite element approximation to eqn (31) becomes

$$\int_{\Omega_{t_n}} (\mathbf{B}_L)_n^T \mathbf{D}(\Delta\boldsymbol{\varepsilon}^E)_{n+1/n}\,d\Omega + (\mathbf{K}_{NL})_n(\Delta\mathbf{U})_{n+1/n} = (\Delta\mathbf{R})_{n+1/n} \quad (34)$$

where $\mathbf{B}_L$ is the linear strain–displacement transformation matrix, $\mathbf{D}$ is the elasticity matrix, $\Delta\boldsymbol{\varepsilon}^E$ is the elastic component of the Green–Lagrange strain increment, $\mathbf{K}_{NL}$ is the non-linear strain incremental stiffness matrix, $\Delta\mathbf{U}$ is the vector of incremental nodal displacements, and $\Delta\mathbf{R}$ is the vector of incremental nodal forces. Considering eqns (33) and (34), each term in eqn (33) is approximated in eqn (34) as

$$\int_{\Omega_{t_n}} C_{ijkl}(\Delta\varepsilon_{kl}^{E})_{n+1/n}(\delta\varepsilon_{ij})_{n+1/n}\,d\Omega \rightarrow \left(\int_{\Omega_{t_n}} (\mathbf{B}_L)_n^T \mathbf{D}(\Delta\boldsymbol{\varepsilon}^E)_{n+1/n}\,d\Omega\right)\Delta\mathbf{U}_{n+1} \quad (35)$$

$$\int_{\Omega_{t_n}} (\sigma_{ij})_n(\delta\eta_{ij})_{n+1/n}\,d\Omega \rightarrow (\mathbf{K}_{NL})_n(\Delta\mathbf{U})_{n+1/n}$$

$$= \left(\int_{\Omega_{t_n}} (\mathbf{B}_{NL})_n^T \boldsymbol{\sigma}^*(\mathbf{B}_{NL})_n\,d\Omega\right)(\Delta\mathbf{U})_{n+1/n} \quad (36)$$

$$(\delta W_e)_{n+1} - \int_{\Omega_{t_n}} (\sigma_{ij})_n(\delta\varepsilon_{ij})_{n+1/n}\,d\Omega = (\Delta\mathbf{R})_{n+1/n}$$

$$= \int_{\Omega_{t_n}} \mathbf{N}^T(\mathbf{b})_{n+1/n}\,d\Omega + \int_{\Gamma_{2_{t_n}}} \mathbf{N}^T(\bar{\mathbf{p}})_{n+1/n}\,d\Gamma - \int_{\Omega_{t_n}} (\mathbf{B}_{NL})_n^T \sigma_n\,d\Omega \quad (37)$$

where $\mathbf{B}_{NL}$ is the non-linear strain–displacement transformation matrix, $\boldsymbol{\sigma}$

is the vector of Cauchy stresses, $\boldsymbol{\sigma}^*$ is a matrix formed by the components of the Cauchy stress tensor, and the remaining quantities have been defined previously.

It is interesting to note that eqn (37) can be recast as

$$
\begin{aligned}
(\Delta \mathbf{R})_{n+1/n} &= \left[\int_{\Omega_{t_n}} \mathbf{N}^{\mathrm{T}} \mathbf{b}_n \, \mathrm{d}\Omega + \int_{\Gamma_{e_{t_n}}} \mathbf{N}^{\mathrm{T}} \bar{\mathbf{p}} \, \mathrm{d}\Gamma - \int_{\Omega_{t_n}} (\mathbf{B}_{\mathrm{L}})_n^{\mathrm{T}} \boldsymbol{\sigma}_n \, \mathrm{d}\Omega \right] \\
&\quad + \left(\int_{\Omega_{t_n}} \mathbf{N}^{\mathrm{T}} (\Delta \mathbf{b})_{n+1/n} \, \mathrm{d}\Omega + \int_{\Gamma_{2_{t_n}}} \mathbf{N}^{\mathrm{T}} (\Delta \bar{\mathbf{p}})_{n+1/n} \, \mathrm{d}\Gamma \right) \\
&= \boldsymbol{\psi}_n + (\Delta \mathbf{f})_{n+1/n}
\end{aligned} \tag{38}
$$

where $\boldsymbol{\psi}_n$ are the out-of-balance (residual) forces at time t_n and $(\Delta \mathbf{f})_{n+1/n}$ the variation of the external forces during the time interval Δt_n ($\Delta t_n = t_{n+1} - t_n$).

4. TIME INTEGRATION ALGORITHMS

In the numerical solution of viscoplastic problems, both explicit and implicit time-stepping algorithms can be employed for integration of the time-rate constitutive equations and the relative merits of the various possible schemes depend upon the physical problem under consideration. Explicit time integration is based on a relatively simple numerical algorithm, but the process is only conditionally stable. Implicit schemes, on the other hand, require more computational effort per time-step but larger time-step lengths can be employed in solution. For the majority of situations an explicit approach provides the most economical results for a given solution accuracy. However, in selected circumstances the use of implicit time integration schemes is either computationally advantageous or unavoidable. For a more detailed discussion on the applicabilities of different integration strategies the reader is referred to Refs 30–32.

In this section, an implicit algorithm proposed[19] for the time integration of the algebraic system represented by eqn (34) is reviewed.

The first step in the development of the solution consists of the evaluation of the elastic component of the linear term of the Green–Lagrange strain increment. From the basic assumption that the strain increments can be decomposed into elastic and viscoplastic parts it follows that

$$
\begin{aligned}
(\Delta \boldsymbol{\varepsilon}^{\mathrm{E}})_{n+1/n} &= (\Delta \boldsymbol{\varepsilon})_{n+1/n} - (\Delta \boldsymbol{\varepsilon}^{\mathrm{vp}})_{n+1/n} \\
&= (\mathbf{B}_{\mathrm{L}})_n (\Delta \mathbf{U})_{n+1/n} - (\Delta \boldsymbol{\varepsilon}^{\mathrm{vp}})_{n+1/n}
\end{aligned} \tag{39}
$$

The viscoplastic strain increment $\Delta\boldsymbol{\varepsilon}^{vp}$ is calculated using an implicit scheme as

$$(\Delta\boldsymbol{\varepsilon}^{vp})_{n+1/n} = (1-\theta)\dot{\boldsymbol{\varepsilon}}_n^{vp} + \theta(\dot{\boldsymbol{\varepsilon}}^{vp})_{n+1/n} \tag{40}$$

where θ, $0 \leq \theta \leq 1$, defines several scheme strategies such as, for example, Euler forward or fully explicit ($\theta = 0$), implicit trapezoidal ($\theta = \frac{1}{2}$), Galerkin ($\theta = 2/3$) and Euler backward or fully implicit ($\theta = 1$).

In order to evaluate the viscoplastic strain-rate terms in eqn (40) it is convenient first to rewrite the constitutive eqns (21) and (22)

$$\dot{\boldsymbol{\varepsilon}}^{vp} = \gamma(\gamma_f, w)\langle\Phi(f, Y)\rangle \frac{\partial F}{\partial\sigma_{ij}} \tag{41}$$

$$\gamma(\gamma_f, w) = \gamma_f/(1-w)^N \tag{42}$$

$$\Phi(f, Y) = \left(\frac{f-Y}{Y}\right)^N \tag{43}$$

The viscoplastic strain rate $(\dot{\boldsymbol{\varepsilon}}^{vp})_{n+1/n}$ is now obtained by using a truncated Taylor series expansion (note that by definition, eqn (31), $\boldsymbol{\sigma}_n \equiv \mathbf{S}_n$), so that

$$(\dot{\boldsymbol{\varepsilon}}^{vp})_{n+1/n} = (\dot{\boldsymbol{\varepsilon}}^{vp})_n + \left(\frac{\partial\dot{\boldsymbol{\varepsilon}}}{\partial\mathbf{S}}\right)_n (\Delta\mathbf{S})_{n+1/n} + \left(\frac{\partial\dot{\boldsymbol{\varepsilon}}}{\partial\gamma}\right)_n (\Delta\gamma)_{n+1/n} + \left(\frac{\partial\dot{\boldsymbol{\varepsilon}}}{\partial Y}\right)_n (\Delta Y)_{n+1/n} \tag{44}$$

Substituting eqn (44) into eqn (40) yields after some algebraic manipulation

$$(\Delta\boldsymbol{\varepsilon}^{vp})_{n+1/n} = \dot{\boldsymbol{\varepsilon}}_n^{vp}\,\Delta t_n + \mathbf{C}_n(\Delta\mathbf{S})_{n+1/n} + \mathbf{E}_n(\Delta\gamma)_{n+1/n} + \mathbf{X}_n(\Delta Y)_{n+1/n} \tag{45}$$

where

$$\mathbf{C}_n = \mathbf{C}_n(\mathbf{S}, \gamma) = \theta\,\Delta t_n \left(\frac{\partial\dot{\boldsymbol{\varepsilon}}}{\partial\mathbf{S}}\right)_n = \theta\,\Delta t_n \mathbf{H}_{1n} \tag{46}$$

$$\mathbf{E}_n = \mathbf{E}_n(\gamma) = \theta\,\Delta t_n \left(\frac{\partial\dot{\boldsymbol{\varepsilon}}}{\partial\gamma}\right)_n = \theta\,\Delta t_n \mathbf{H}_{2n} \tag{47}$$

$$\mathbf{X}_n = \mathbf{X}_n(\mathbf{S}, \gamma, Y) = \theta\,\Delta t_n \left(\frac{\partial\dot{\boldsymbol{\varepsilon}}}{\partial Y}\right)_n = \theta\,\Delta t_n \mathbf{H}_{3n} \tag{48}$$

The gradient matrices $\mathbf{H}_1$, $\mathbf{H}_2$ and $\mathbf{H}_3$ are obtained from eqns (41)–(43) as

$$\mathbf{H}_{1n} = \gamma_n \left\{ \phi_n \left[\frac{\partial}{\partial \mathbf{S}} \left(\frac{\partial F}{\partial \mathbf{S}} \right)_n \right]_n + \left(\frac{\partial \phi}{\partial \mathbf{S}} \right)_n \left(\frac{\partial F}{\partial \mathbf{S}} \right)_n^{\mathrm{T}} \right\} \tag{49}$$

$$\mathbf{H}_{2n} = \phi_n \left(\frac{\partial F}{\partial \mathbf{S}} \right)_n \tag{50}$$

$$\mathbf{H}_{3n} = \gamma_n \left(\frac{\partial \phi}{\partial Y} \right)_n \left(\frac{\partial F}{\partial \mathbf{S}} \right)_n \tag{51}$$

The second Piola–Kirchhoff stress increment is obtained from eqn (32) to be

$$\Delta \mathbf{S}_{n+1/n} = \mathbf{D}(\Delta \boldsymbol{\varepsilon}^{\mathrm{E}})_{n+1/n} \tag{52}$$

Substituting eqn (45) into eqn (39) and using eqn (52) leads to

$$(\Delta \boldsymbol{\varepsilon}^{\mathrm{E}})_{n+1/n} = (\mathbf{I} + \mathbf{C}_n D)^{-1} \times [(\mathbf{B}_{\mathrm{L}})_n (\Delta \mathbf{U})_{n+1/n} - \dot{\boldsymbol{\varepsilon}}_n^{\mathrm{vp}} \Delta t_n - \mathbf{E}_n (\Delta \gamma)_{n+1/n} - \mathbf{X}_n (\Delta Y)_{n+1/n}] \tag{53}$$

where $\mathbf{I}$ is the identity matrix of order 6.

Introducing eqn (53) into the incremental equilibrium equation (34) and solving for the displacement increment results in the following equation system

$$\Delta \mathbf{U}_{n+1/n} = (\mathbf{K}_{\mathrm{T}})_n^{-1} \Delta \mathbf{V}_{n+1/n} \tag{54}$$

where $\mathbf{K}_{\mathrm{T}}$, termed the tangential stiffness, is given by

$$(\mathbf{K}_{\mathrm{T}})_n = \int_{\Omega_{t_n}} (\mathbf{B}_{\mathrm{L}})_n^{\mathrm{T}} \hat{\mathbf{D}}_n (\mathbf{B}_{\mathrm{L}})_n \, \mathrm{d}\Omega + (\mathbf{K}_{\mathrm{NL}})_n \tag{55}$$

in which $\hat{\mathbf{D}} = (\mathbf{D}^{-1} + \mathbf{C}_n)^{-1}$ and $\Delta \mathbf{V}_{n+1/n}$, termed the incremental pseudo-load vector, is given by

$$\Delta \mathbf{V}_{n+1/n} = \Delta \mathbf{R}_{n+1/n} + \int_{\Omega_{t_n}} (\mathbf{B}_{\mathrm{L}})^{\mathrm{T}} \hat{\mathbf{D}}_n (\dot{\boldsymbol{\varepsilon}}_n^{\mathrm{vp}} \Delta t_n + \mathbf{E}_n \Delta \gamma_{n+1/n} + \mathbf{X}_n \Delta Y_{n+1/n}) \, \mathrm{d}\Omega \tag{56}$$

Equation (54) cannot be solved directly because the increments of the γ and Y functions are as yet unknown. Estimates of these increments are now discussed.

Considering eqn (42), $\Delta \gamma_{n+1/n}$ is readily obtained as

$$\Delta \gamma_{n+1/n} = N \gamma_{\mathrm{f}} \, \Delta w_{n+1/n} / (1 - w_n)^{N+1} \tag{57}$$

where the increment of damage $\Delta w_{n+1/n}$ is predicted using constitutive eqn (22) and an Euler forward scheme as

$$\Delta w_{n+1/n} = A \frac{(\hat{\sigma}_{eq})_n^{\chi}}{(1 - w_n)^{\phi}} \Delta t_n \tag{58}$$

The increment of the Y function, on the other hand, is obtained from eqn (23) as

$$\Delta Y_{n+1/n} = H' \, \Delta \bar{\varepsilon}^{vp}_{n+1/n} \tag{59}$$

where the increment of the effective viscoplastic strain component $\Delta \bar{\varepsilon}^{vp}_{n+1/n}$ is predicted from constitutive eqn (21) and an Euler forward scheme as

$$\Delta \bar{\varepsilon}^{vp}_{n+1/n} = \dot{\bar{\varepsilon}}_n \, \Delta t_n \tag{60}$$

With the evaluation of the pseudo forces complete, the displacement increments are finally found from eqn (54) and subsequently the strain increments from eqns (45) and (39), and the second Piola–Kirchhoff stress increments from eqn (52). With these increments known, the displacements, strains and stresses (second Piola–Kirchhoff and Cauchy) are updated to their values at time t_{n+1}. Updating of the damage state variable is carried out by the following predictor–corrector scheme where the superscripts 'p', 'c' and 'a' denote predicted, corrected and accepted values, respectively

$$w^{p}_{n+1/n} = w_n + \frac{A(\hat{\sigma}_{eq})_n^{\chi}}{(1 - w_n)^{\phi}} \Delta t_n \tag{61}$$

$$w^{c}_{n+1/n} = \frac{1}{2}\left\{(w^{p}_{n+1/n} - w_n) + \frac{A(\hat{\sigma}_{eq})_{n+1}^{\chi}}{(1 - w^{p}_{n+1/n})^{\phi}}\right\} \Delta t_n \tag{62}$$

$$w^{a}_{n+1} = \tfrac{1}{5}(w^{p}_{n+1/n}) + \tfrac{4}{5}(w^{c}_{n+1/n}) \tag{63}$$

The updated values at time t_{n+1} are accepted and the solution advanced to the next time-station provided that two accuracy controls, discussed below, are satisfied. Rupture of a particular point in the structure is deemed to have occurred when the damage state variable reaches the critical value 1 (in practice, for numerical convenience, the critical value is taken as 0·99). The residual stress existing at this point at the instant it is declared ruptured is automatically redistributed to the remaining non-ruptured elements under the assumption that the real redistribution process is both instantaneous and elastic. The analysis proceeds until a sufficiently large damaged zone develops characterising the global failure of the structure.

The first rule to select the magnitude of the time-step consists in limiting the maximum amount of incremental viscoplastic strain to a fraction of the total accumulated strain according to

$$\Delta t_{n+1} = \min\,(\tau \bar{\varepsilon}_n / \dot{\bar{\varepsilon}}_n^{\mathrm{vp}}) \tag{64}$$

where τ is a pre-specified value, and $\bar{\varepsilon}$ and $\dot{\bar{\varepsilon}}^{\mathrm{vp}}$ are the effective values of the total strain and of the viscoplastic strain rate, respectively. The minimum value of Δt_{n+1} is taken with respect to all integrating points in the domain. The second rule limits the change in the time-step length between any two intervals, i.e.

$$\Delta t_{n+1} \leq k\, \Delta t_n \tag{65}$$

where k is a given constant. Numerical experience to date indicates that suitable values for τ and k are in the ranges $0{\cdot}01 \leq \tau \leq 0{\cdot}15$ and $1{\cdot}5 \leq k \leq 3{\cdot}0$, respectively, although there are no fixed criteria for their specification.

Stability and accuracy of the solution are controlled by requiring that at any time-step the Cauchy stress variation and the error in the damage increment prediction at each integrating point should not exceed pre-specified tolerances, i.e.

$$(1) \left| \frac{\|\boldsymbol{\sigma}_{n+1}\| - \|\boldsymbol{\sigma}_n\|}{\|\boldsymbol{\sigma}_n\|} \right| \leq \mathrm{toler}_1 \tag{66}$$

$$(2) \left| \frac{\Delta w^{\mathrm{a}}_{n+1/n} - \Delta w^{\mathrm{p}}_{n+1/n}}{\Delta w^{\mathrm{p}}_{n+1/n}} \right| \leq \mathrm{toler}_2, \qquad w_n \geq 0{\cdot}1 \tag{67}$$

where $\| \ \|$ denotes the usual Euclidean norm, $| \ |$ the absolute value, and toler_1 and toler_2 the accuracy tolerances. If either condition (1) or condition (2) is violated at any integrating point, the current time-step length is reduced to $\frac{1}{5}$th of its value and the time-step solution repeated. A maximum of two reanalyses is allowed for any time-step after which, if the accuracy controls are still not satisfied (usually a rare situation if τ and k are carefully chosen) the solution is halted and the last calculated results stored on file. This procedure gives the analyst an opportunity to examine the deformation and damage histories and to decide whether to restart the analysis from the current solution at time t_n adopting a time-step length smaller than the one automatically calculated or, if restart facilities are provided, from any of the previous time-station solutions stored.

5. NUMERICAL APPLICATION

In order to illustrate the applicability of the technique, the creep damage and rupture behaviour of a pressure vessel under constant load are investigated.

The vessel considered is of spherical form with a reinforced cylindrical nozzle (Fig. 1) and is made of commercially pure copper (BS 2873–CIDI). The geometric data chosen are: $R = 66$ mm, $T = 1{\cdot}73$ mm, $r = 19{\cdot}8$ mm, $t = 1{\cdot}73$ mm, and $c = 2{\cdot}92$ mm. All relevant uniaxial material data for the copper are compiled in Table 1 and correspond to a temperature of 250 ± 1 °C. The material constant M, appearing in Table 1, refers to a polynomial term in time, t^M, which has been included in the creep strain rate and damage rate equations in order to account for primary creep

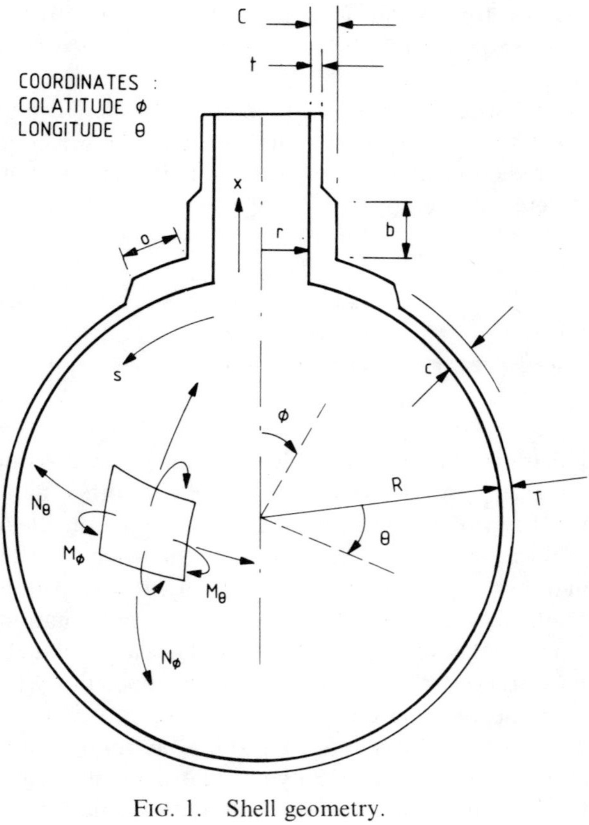

FIG. 1. Shell geometry.

effects.[33] This polynomial term can be most conveniently dealt with by introducing a new time-scale t^*, defined by $t^* = t^{(M+1)}$. When recast in terms of t^*, the creep strain rate and the damage rate equations become independent of t^* but the A and γ_f constants now appear multiplied by the factor $1/(M+1)$.

As previously mentioned, the rupture behaviour of commercially pure copper is closely approximated by the maximum principal stress rupture criteria. For the particular copper used, the values of the α_i parameters (eqn (20)) are: $\alpha_1 = 0{\cdot}848$, $\alpha_2 = 0{\cdot}0$, $\alpha_3 = 0{\cdot}064$ and $\alpha_4 = 0{\cdot}088$.

In the solution procedure, advantage was taken of the axisymmetric

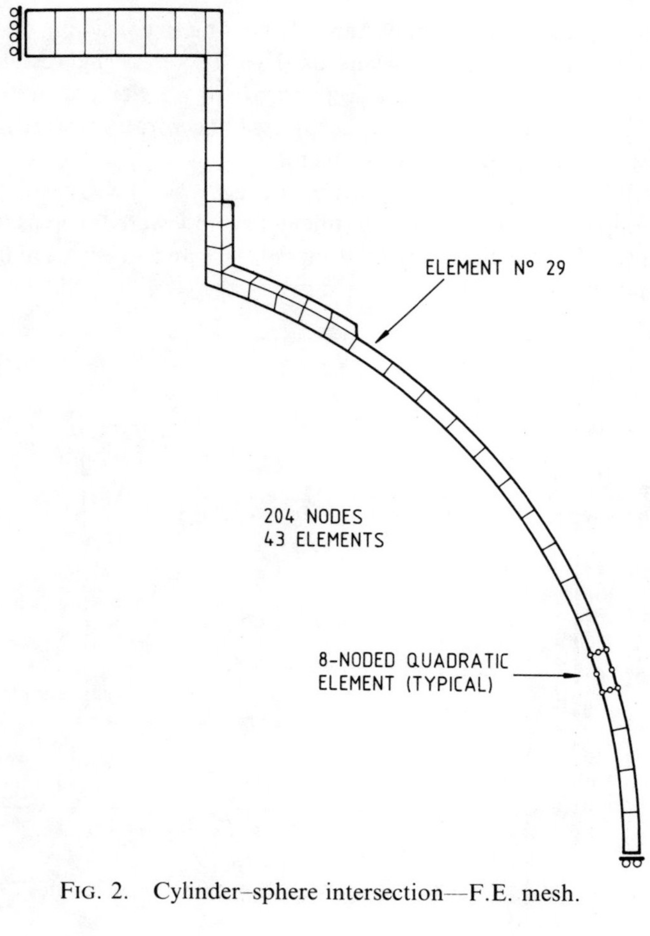

FIG. 2. Cylinder–sphere intersection—F.E. mesh.

TABLE 1
MATERIAL PROPERTIES

Property	*Numerical value*
E	$66\,240{\cdot}0\ \mathrm{N/mm^2}$
ν	$0{\cdot}3$
γ_f	$3{\cdot}21 \times 10^{-12}/\mathrm{h}$
N	$5{\cdot}0$
A	$1{\cdot}89 \times 10^{-7}\,(\mathrm{N/mm^2})^{-5{\cdot}0}/\mathrm{h}$
χ	$3{\cdot}19$
ϕ	$6{\cdot}0$
M	$-0{\cdot}43$

characteristics of the problem. A finite element mesh consisting of 43 eight-noded isoparametric elements was used to discretise the vessel, and is illustrated in Fig. 2. The distribution and size of the elements were chosen so that the expected regions of higher stress concentrations were modelled by a greater number of smaller elements.

The implicit–trapezoidal integration strategy ($\theta = \frac{1}{2}$) was employed and the time-step length controlling parameters τ and k were taken as 0·05 and 2·0, respectively. The accuracy tolerances $toler_1$ and $toler_2$ were 0·10 and 0·05, respectively.

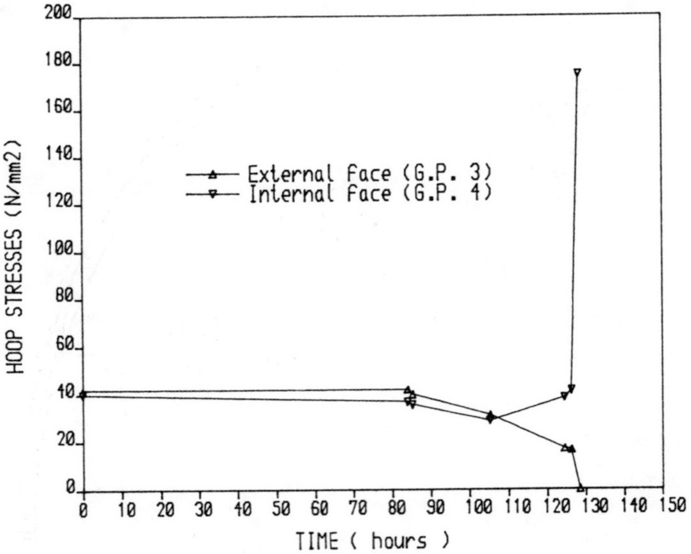

FIG. 3. Variation of hoop stresses with time at element no. 29.

For a constant applied pressure of 2·0 N/mm^2 the computed rupture time was ~128·5 h. For the particular geometry, material and load data combination chosen, the effects of geometry changes on the vessel damage and rupture behaviour were not pronounced, as indicated by the rupture time of ~127·5 h obtained in a geometric linear analysis.

Figure 3 shows the variation of the hoop stresses at the internal and external faces of element no. 29. It can be seen that for most of the vessel lifetime (~70 %) the distribution of hoop stresses remains constant. At the

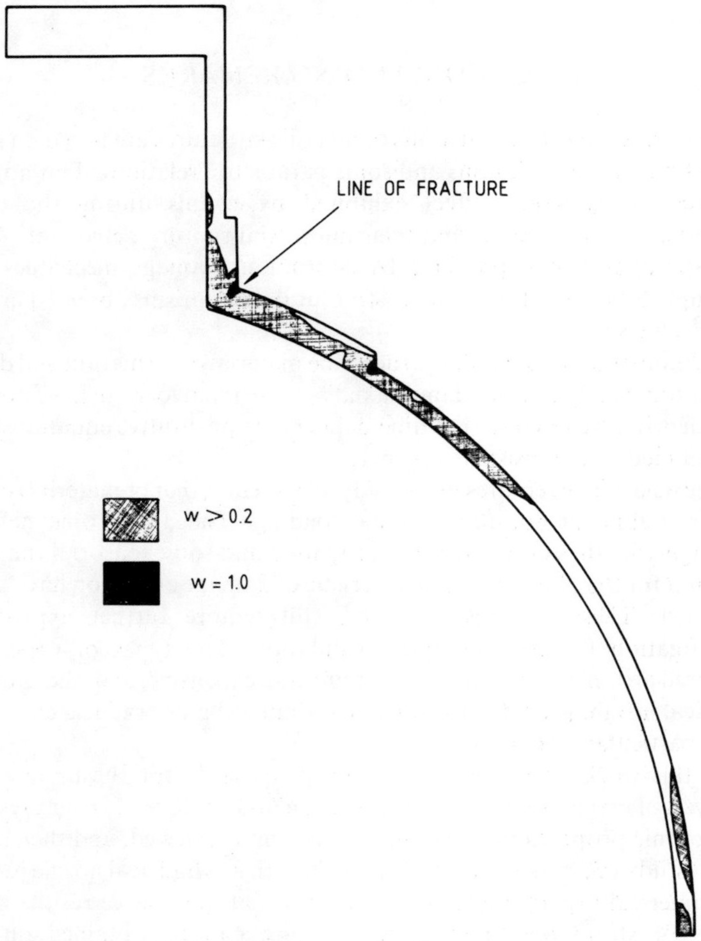

FIG. 4. Damage contours at time $t = 128{\cdot}5$ h.

onset of tertiary creep a strong stress redistribution occurs with the crack front moving from the outer surface inwards.

Two zones of computed damage are presented in Fig. 4 at a time immediately prior to total collapse. The first region, with $w \geq 0{\cdot}2$, corresponds to moderately damaged material while the second, with $w = 1{\cdot}0$, represents failed material. As can be observed, the main failure zone is close to the cylinder–sphere intersection and the vessel failed by the cylindrical portion of the nozzle shearing away from the main part of the sphere.

6. CONCLUDING REMARKS

The classical lifetime predictions of metallic structures are based on steady-state stress–strain solutions and some parametric relations. This approach ignores the softening effect exhibited by metals during the tertiary creep/viscoplastic regime and their multiaxial rupture behaviour. A more realistic approach is provided by continuous damage mechanics which attempts to account for the microstructural mechanisms observed in metals under stress.

Phenomenological models to describe material deformation and damage under uniaxial loading and multiaxial proportional loading have now been sufficiently developed so that time-dependent constitutive equations can be determined in a consistent manner.

The field which requires most study at present is that of material response under multiaxial non-proportional loading. The anisotropic nature of damage growth is now generally recognised and some tensorial theories to account for the directional characteristic of damage evolution have already emerged. These theories, however, still require further experimental investigation for their complete validation. The types of experiments required are, in general, time-consuming and expensive, and, therefore, any physical insight provided by numerical simulation of practical case studies is of particular importance.

In this work an implicit algorithm proposed[19] for the finite element analysis of creep/viscoplastic deformation and rupture of structures under monotonic proportional loading has been first reviewed, and then applied to the study of a spherical pressure vessel with a cylindrical nozzle junction.

Numerical experience to date indicates that qualitative results both in terms of strain and damage distribution can be obtained and that reasonable estimates of time to rupture (considering the highly non-linear nature of the problem and the scatter normally associated with the material data) are possible for the class of problems considered. In common with most types of non-linear analysis, the demands on computer resources are high, especially when dealing with large and complex structures.

REFERENCES

1. Hill, R. (1950). *The Mathematical Theory of Plasticity*, Oxford University Press, Oxford.
2. Perzyna, P. (1963). *Quarterly Applied Mathematics*, **20**, 321–32.
3. Perzyna, P. (1966). *Fundamental Problems in Viscoplasticity*, *Recent Advances in Applied Mechanics*, Academic Press, New York, Vol. 9, pp. 243–377.
4. Olszak, W. and Perzyna, P. (1966). On elastic-viscoplastic soils, rheology and soil mechanics, *IUTAM Symposium*, Springer-Verlag, Grenoble.
5. Olszak, W. and Perzyna, P. (1970). In: *Inelastic Behaviour of Solids*, (eds M. F. Kanninen *et al.*), McGraw-Hill, New York.
6. Owen, D. R. J., Prakash, A. and Zienkiewicz, O. C. (1974). *Computers & Structures*, **4**, 1251–67.
7. Kachanov, L. M. (1959). *Izv. Akad. Nauk. USSR, Otdgel, Tekh. Nauk.*, **8**, 26–31.
8. Rabotnov, Y. N. (1968). Creep rupture, *Proc. XII Int. Cong. Appl. Mech.*, Stanford.
9. Hult, J. (1979). *CDM—Capabilities, Limitations and Promises—Mechanisms of Deformation and Fracture*, Pergamon Press, Oxford, 233–47.
10. Chaboche, J. L. (1981). *Nuclear Engng. and Design*, **64**, 233–47.
11. Leckie, F. A. and Hayhurst, D. R. (1974). *Proc. R. Soc., London*, **A340**, 323–47.
12. Sidoroff, F. (1980). *Proc. IUTAM Symposium on Physical Nonlinearities in Structural Analysis*, Springer-Verlag, Berlin, 237–43.
13. Murakami, S. and Ohno, N. (1978). *IUTAM Colloquium Euromech 111*, Marienbad.
14. Chaboche, J. L. (1979). *Colloquium Euromech 115*, Grenoble.
15. Bodner, S. R. (1981). *Proc. IUTAM Symposium on Physical Nonlinearities in Structural Analysis*, Springer-Verlag, Berlin, 21–8.
16. Betten, J. (1981). *Euromech Colloquium 147*, *Damage Mechanics*, Cachan.
17. Murakami, S. and Imaizumi, T. (1982). *J. Mecanique Theorique et Appliquée*, **1** (5).
18. Hayhurst, D. R., Trampczynski, W. A. and Leckie, F. A. (1980). *Acta Metallurgica*, **28**, 1171–983.
19. Gonçalves F^{O}., O. J. A. and Owen, D. R. J. (1983). In: *Engineering Approaches to High Temperature Design*, (eds B. Wilshire *et al.*), Pineridge Press, Swansea.
20. Bingham, E. D. (1922). *Fluidity and Plasticity*, McGraw-Hill, New York, pp. 215–18.
21. Reiner, M. (1955). *Rheologie Theorique*, Dunod, Paris.
22. Freudenthal, A. M. and Geiringer, H. (1958). *Encyclopedia of Physics*, **6**, 229–433.

23. Johnson, A. E., Henderson, J. and Khan, B. (1962). *Complex Stress Creep, Relaxation and Fracture of Metallic Alloys*, HMSO, London.
24. Owen, D. R. J. and Hinton, E. (1980). *Finite Elements in Plasticity: Theory and Practice*, Pineridge Press, Swansea.
25. Gallagher, R. H. (1975). *Finite Element Analysis—Fundamentals*. Prentice-Hall, Englewood Cliffs.
26. Leckie, F. A. (1982). *Int. J. Mech. Sci.*, **24**(4), 245–50.
27. Bathe, K. J., Ramm, E. and Wilson, E. L. (1975). *Int. J. Num. Meth. Engng.*, **9**, 353–86.
28. Fung, Y. C. (1965). *Foundations of Solid Mechanics*, Prentice-Hall, Englewood Cliffs.
29. Malvern, L. E. (1969). *Introduction to the Mechanics of a Continuum Medium*, Prentice-Hall, Englewood Cliffs.
30. Hayhurst, D. R. and Krzeczkowski, K. (1979). *Comp. Meth. Appl. Mech. Engng.*, **20**, 151–71.
31. Kumar, V., Morjaria, M. and Mukherjee, S. (1980). *ASME, J. Engng. Materials and Technology*, **102**, 92–6.
32. Argyris, J. H., Vaz, L. E. and Willam, K. J. (1978). *Comp. Meth. Appl. Mech. Engng.*, **16**, 231–77.
33. Hayhurst, D. R., Dimmer, P. R. and Chernuka, M. W. (1975). *J. Mech. Phys. Solids*, **23**, 335–55.

6

Property Interactions, Multiaxial Stressing and Component Testing

F. SCHUBERT and H. J. PENKALLA
Institut für Reaktorwerkstoffe,
Kernforschungsanlage Jülich GmbH, Jülich,
Federal Republic of Germany

H. WEBER
Mannesmann–Forschungsinstitut GmbH, Duisburg,
Federal Republic of Germany

and

K. BIENIUSSA
Gesellschaft für Reaktorsicherheit, Köln,
Federal Republic of Germany

1. INTRODUCTION

1.1. General Introductory Remarks

In the previous contributions to this seminar, rules and methods for the dimensioning, construction and analysis of component behaviour have been discussed and some information concerning the materials data and constitutive equations needed has been given. This contribution discusses the failure modes which can occur in high temperature metallic components due to long-term loading and the interactions of different loading conditions dependent upon operating history. The instructions given in design standards for metallic components, in the temperature range where time-dependent properties dominate for design configuration and for strength and strain evaluation, will be examined using available experimental results. The experiences are based on the results of the material evaluation programmes for advanced high temperature gas cooled reactors,[1] on more than three years effort in Germany to develop a structural design standard for very high operating temperatures,[2] on results on gas turbine materials,[3,4] and on knowledge of the behaviour of long-term exposed components of non-nuclear power stations.[5] The discussion provides examples, which may not be generalised for all high temperature materials. The long-term behaviour of high temperature alloys is highly dependent on the microstructure and on the stability of the microstructure at the operating temperature; this aspect of materials behaviour should be included in the design considerations.

High temperature components can be designed by rules or by analysis:

(a) '*Design by rules*' is by following regulations such as the German AD Instructions[6] and TRD proposals,[7,8] mainly used for non-nuclear pressure vessels and boilers; design is undertaken against creep rupture due to mechanical loading. The necessary allowable stresses are specified using average creep and creep-rupture data. Other time-dependent failure modes, e.g. creep-fatigue failure, are not adequately considered. Components designed by these standards have been in operation in non-nuclear power stations for more than 100 000 h and now the estimation of the disposable remaining life is becoming important.

(b) '*Design by analysis*': for an adequate design, an expected loading history is supposed, and then an attempt is made to make an analysis of the component behaviour. A differentiation between 'primary stresses' and 'secondary stresses' must be thoroughly evaluated and the interaction calculated. This is the approach adopted in design codes such as the KTA rules[9] or ASME codes;[10] only ASME Code Case N47[11] extends into the temperature range where time-dependent failure modes occur.

The aim of both approaches is to avoid catastrophic failure, as demonstrated for example in Fig. 1 which shows a tube which failed within the tube elbow after a 46 000-h service life. The reason for the spontaneous failure has been an extended area of creep damage nearly through the cross-section in the tension zone of the tube. This locally damaged zone led to an instability of the cross-section and represented the starting flaw for spontaneous crack growth. This zone is not a critical crack size, as it is used in elastic-plastic fracture mechanics analysis, but an area with a lot of creep voids and creep cracks through the wall, possibly resulting from a superposed external additional stress due to forces in the pipe system. The

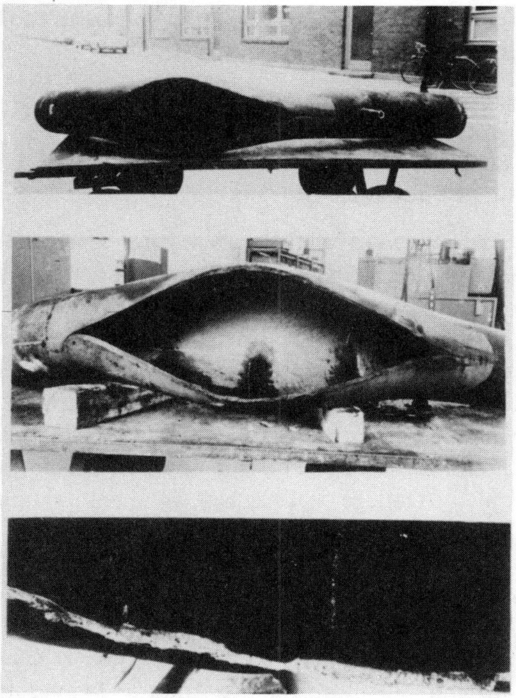

FIG. 1. Burst tube bend after creep damage ($T = 540\,°C$, time $= 46\,000$ h).

relatively short-term break after 40 000 h was unexpected and not predictable during the last inspection period.

1.2. Definition of Loading

A component such as a tube may be exposed to stresses due to the pressure differences across the wall thickness and its deadweight. These stresses are in equilibrium with the loading conditions and are defined as primary stresses which exhaust the material by creep.

Start-up and shut-down procedures and power adjustment effects lead to a change in the strain range (material exhaustion due to fatigue) and in the stress level (material exhaustion due to creep) and, in this way, to a combined loading condition. For high temperature applications the determination of strain range and stress level using elastic analysis can lead to difficulties on account of conservative and unrealistic stress levels. Relaxation and/or creep procedures alter the stress distribution in a component which is determined by inelastic analysis. For the inelastic analysis with finite-element methods, the following are needed for the range of operating temperature:

(a) Stress–strain–time curves.
(b) Stress rupture curves.
(c) Creep–fatigue interaction criteria.
(d) Constitutive equations.

The use of data and rules, derived from results of uniaxial tested specimens, must be clarified for describing the behaviour under multiaxial loading conditions.

In addition to the operating parameters, the material is additionally exposed to fault conditions or alterations in the pressure and/or temperature levels. The existence of flaws or areas of imperfect microstructure in a component may have a significant influence on its lifetime. Welds and heat affected zones in components normally exhibit a certain susceptibility to failure occurrences; weldments should be placed in the component outside the highest loading conditions. The welding procedure as well as the post-welding heat tratment must be optimised for the material used. In stress analysis, the special properties across the joint should be recognised.

At high operating temperatures, isothermal exposure causes instability in microstructure and, therefore, the duration at temperature has to be considered if short-term mechanical properties after long-term application are of importance for the component behaviour.

1.3. Definition of Failure Modes and Kinds of Interaction

For metallic components, operating at high temperatures, the following failure modes must be discussed:

(a) Creep rupture due to long-term loading.
(b) Creep-fatigue failure due to cycles in the operating conditions.
(c) Gross distortion due to incremental collapse and ratcheting caused by cyclic loading.
(d) Loss of function due to excessive deformation.
(e) Creep buckling.
(f) Failure due to environmental attack.

In practice, failure can be a combination of different failure modes. The high temperature loading conditions which have a long-term influence are:

(a) Constant stress and alternating stresses causing creep.

(b) Relaxation causing deformation and damage.
(c) Strain cycling causing fatigue (LCF).
(d) Stress cycling causing fatigue (HCF).
(e) Corrosion causing surface defects.

and, in the case of nuclear components:

(f) Neutron irradiation causing embrittlement.

The results of these loading conditions are a change in microstructure, and microstructural damage of the material and of the surface of the component. The macroscopic results are deformation, surface scaling, flaws or cracks, and fracture. The effect of each kind of loading is dependent on the loading history. A separation of the effect of each kind on the component behaviour is an approximation. For example, dependent on the local gas/metal reactions at the surface of the material, corrosion can influence mechanical properties, crack initiation, crack propagation and failure.

2. CONSIDERATION OF CREEP

2.1. Creep, Relaxation and Damage

A simple scheme of 'design by rules' is given in Fig. 2. The stress criteria define the limitation for the mechanical loads. The stresses resulting from

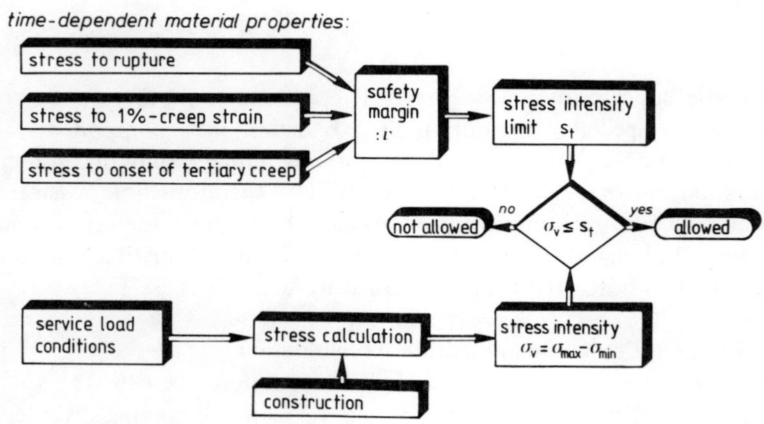

FIG. 2. Flow diagram for stress limitation at temperatures evaluated according to ASME CC N47.

the multiaxial loading conditions are combined to an effective stress (stress intensity) and then compared with the stress intensity limit (see Chapter 1).

The calculation and comparison of the stresses include secondary stresses which are induced by temperature transients. Secondary stresses are self-limiting. At lower temperatures the limit is given by the yield strength, whereas at high temperatures the limit is time-dependent and decreases with time, The effect of stress relaxation is creep-controlled. At higher temperatures the induced stresses relax more quickly than at lower temperatures. At higher temperatures the secondary stresses may be of minor importance.

Time-dependent strain and relaxation are described by constitutive equations which are the basis of inelastic analysis (see Chapters 3 and 5). In inelastic analysis Norton's creep law[12]

$$\dot{\varepsilon} \sim \sigma^n$$

where $\dot{\varepsilon}$ is the strain rate, σ is the stress and the superscript n is a materials parameter, has often been used to calculate the stress–strain behaviour.

Creep and relaxation lead to material damage. The increased plastic strain initiates the growth of voids and creep cracks in the material. The incubation time for the initiation of cracks and voids depends on material structure, stress and temperature.

In recent theories the damage D of material is defined as the ratio of the cross-section of voids A_c to the total measurable cross-section A. To describe the time dependence of creep damage Kachanov's[13] equation is used

$$D \sim \sigma^p$$

where the superscript p is a materials parameter. The creep damage represents a special contribution to exhaustion in a component. The correlation between the amount of material damage and the grade of exhaustion is not yet fully understood. It is therefore not possible to estimate the residual lifetime of a component by the local material damage situation. Exhaustion is the consumption of the permissible operating time, which is both stress- and temperature-dependent.

2.2. Lifetime Prediction and Strain Accumulation

In order to predict component failure, at different creep stresses, an equation for creep damage is needed. The creep life at changing stress levels is usually derived from the well-known behaviour under single stress at different test temperatures. With simple rules, individual time-steps with constant stress or simplified constant loads are summed. Robinson's[14] rules

$$\sum \frac{t}{t_m} = 1$$

where t is time and t_m is time to rupture at a given stress level, or those of Mordfin *et al.*[15]

$$\sum \frac{t}{t_m} = L_m$$

are very common. Creep fracture has to be expected when the accumulation of lifetime fractions reaches either a value of 1 or the medium relative creep rupture lifetime fraction L_m. The former is proposed in design rules for conventional components (e.g. TRD 508).

Only restricted experimental results of stress change tests with practical stress and strain levels are reported.[16,17] There exist results on specimens of steam boilers and steam turbine steels which were tested under rectangular cyclic stresses. In Fig. 3 these test results are compared to calculations via lifetime fraction rules. Up to about 30 000 h, the lifetime fraction rule predicts the lifetime with sufficient precision. For stress cycling, a value of 0·7 has been found, and for temperature cycling a value of 0·92 has been established.[16]

A linear accumulation rule satisfactorily describes the creep-rupture behaviour of the Ni-based alloy IN 713 LC under different stress levels at 850 °C (Figs 4 and 5).[17]

The creep curve of a specimen under alternating creep load can be composed by adding parts of individual creep curves, obtained at those stresses and temperature conditions, which are representative for each loading condition (schematic examples are given in Fig. 6).

Under the loading condition σ_1, ϑ_1 (where ϑ_1 is a given temperature) after a time t_e a creep deformation is reached according to the creep curve up to point e. Beyond point e the creep curve does no longer follow the original curve defined by the loading condition σ_1, ϑ_1, but changes because of the higher loading condition σ_2, ϑ_2. It is expected that processing deformation is according to the creep curve σ_2, ϑ_2; however, definitions are not given of where to start and what part of the creep curve σ_2, ϑ_2 has to be added. Three hypotheses are helpful in making these decisions:

(a) The time-hardening rule takes into account the influence of exposure time at a given stress level.

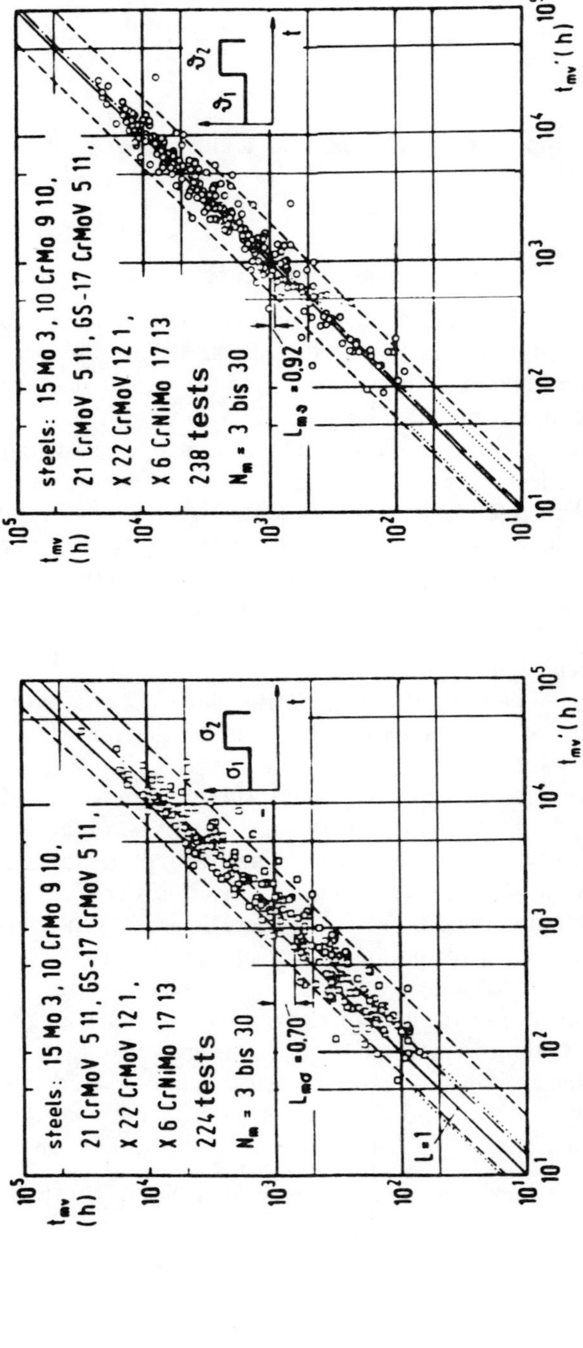

FIG. 3. Comparison of creep results with rectangular cycling stress and lifetime calculations.[16] (N_m is number of cycles to fracture, $L_{m\sigma}$ is medium lifetime fraction at stress σ_L, t_{mv} is experimental time to rupture.)

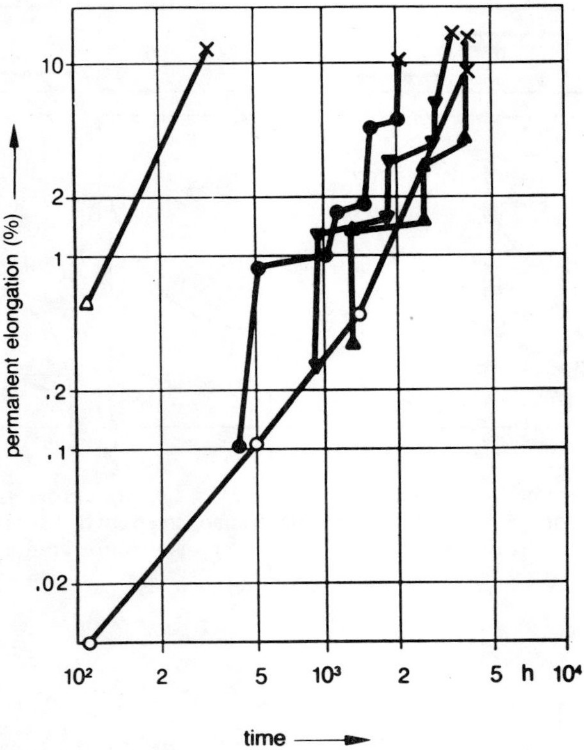

FIG. 4. Creep-rupture cycling test of IN 713 LC at 850 °C.[17]

		Peak fraction X (%)
○ 163 N/mm²	● 163–273 N/mm²	7·7
△ 273 N/mm²	▲ 163–273 N/mm²	2·6
	▼ 163–383 N/mm²	0·3

(b) The strain-hardening rule considers the level of the given remaining strain.

(c) The energy-hardening rule accounts for the consumed volume deformation energy.

All these results have been obtained on uniaxially loaded specimens.

Using Kachanov's[13] equations and Norton's creep law[12] a system of coupled differential equations[18] is obtained, which can be used to calculate the lifetime under alternating loads more precisely. In Fig. 7 experimental results on Inconel 617 are fitted by these equations. For the extrapolation of stress to rupture curves they can be used as long as Norton's law is valid.

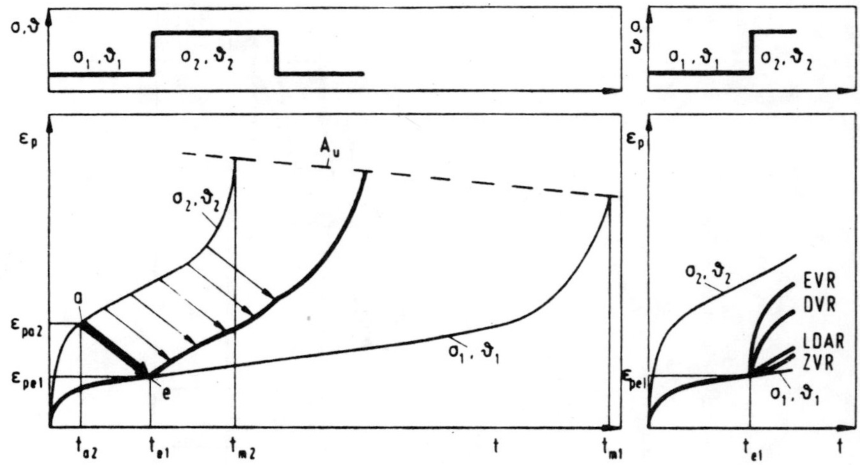

FIG. 5. Creep behaviour under cyclic loading.[16] (A_u is rupture deformation, ε_{pe1} is plastic strain of creep curve 1, ε_{pa2} is plastic strain of creep curve 2, t_{e1} is time of curve 1 to point e, t_{a2} is time of curve 2 to point a, t_{m2} is time to rupture of curve 2.)

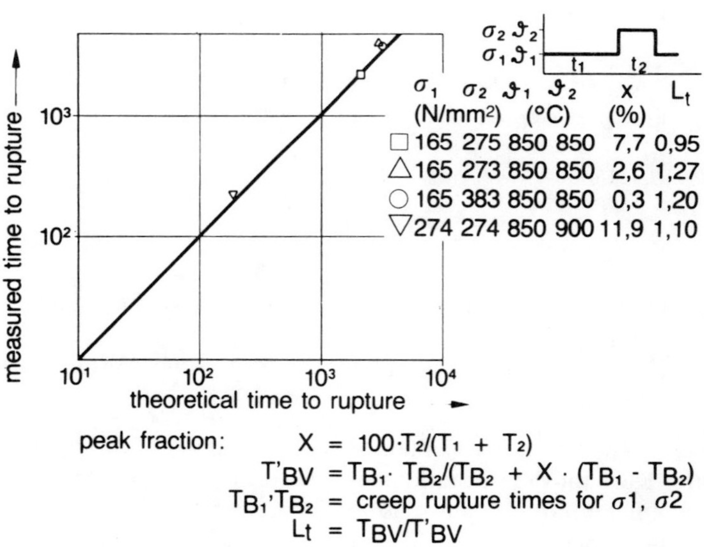

FIG. 6. Comparison of damage accumulation rule and test results on IN 713 LC.[17]

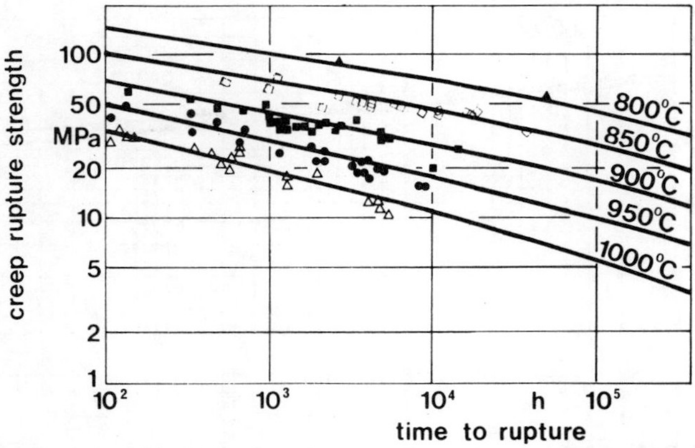

FIG. 7. Comparison of test results on Inconel 617 with the calculated medium creep rupture curve.[18]

To describe the creep behaviour, similar ideas may have led to the RWTÜV proposal[19] whereby the remaining allowable operating time is estimated by the evaluation of surface replicas of components (Fig. 8). This method has to be treated with caution; there is no established quantitative relationship between exhaustion and damage since no experimental results exist which indicate that the degrees of exhaustion and of damage have the

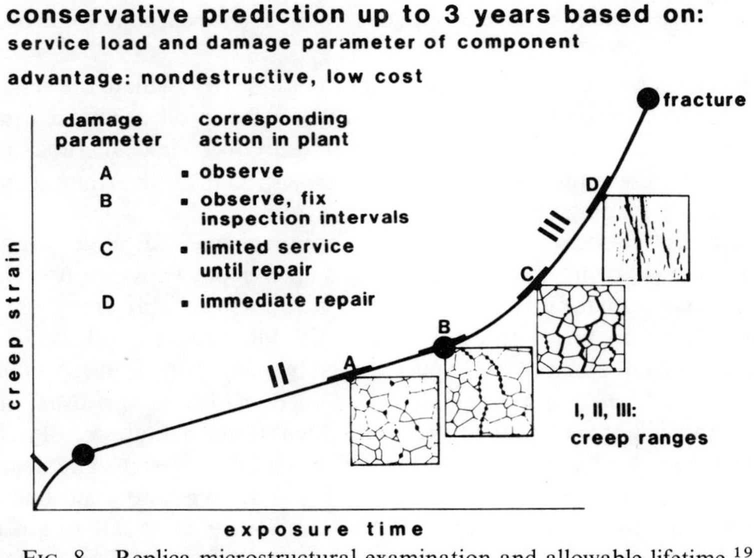

FIG. 8. Replica microstructural examination and allowable lifetime.[19]

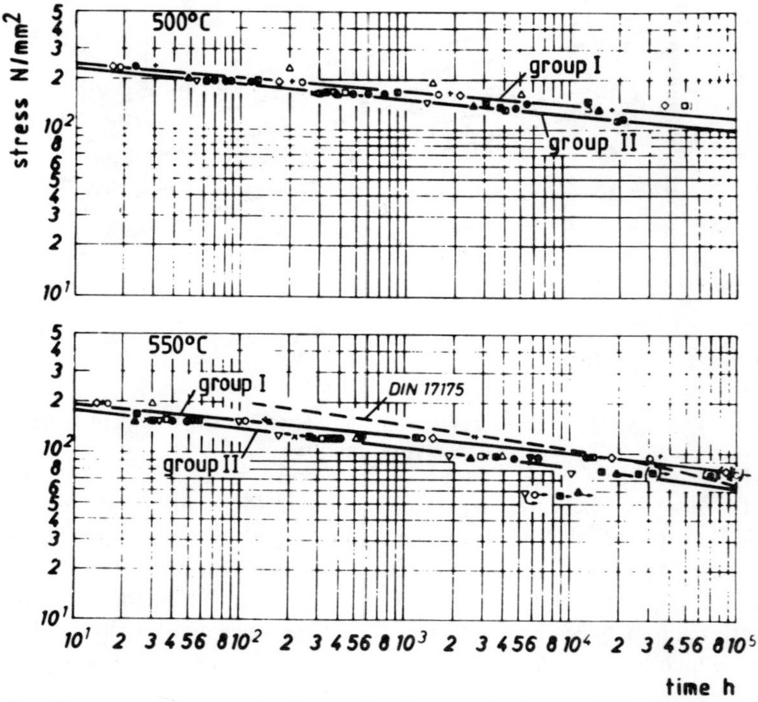

FIG. 9. Creep-rupture strength of operational exposed components of 10CrMo9 10.

same time dependence. It is therefore not possible to calculate the residual lifetime of a component from the state of the surface replica. The treatment of these surface microscopic flaws with fracture mechanics methods may lead to wrong remanent lifetime predictions, especially if the component is under creep-loading conditions.

That this treatment has to be carefully approached is indicated by results (Figs 9 and 10) of creep tests on specimens taken from walls of tubes which were in operation for up to 100 000 h.[20] The tubes were fabricated from the high temperature resistant ferritic steel 10 Cr–Mo 9 10. Group I specimens have a special microstructure, characterised by a Mo_2C-pearlite; specimens of Group II are normalised and air-cooled, which is the usual heat treatment. The creep-rupture time of specimens tested at high stress levels is below the scatter band of DIN 17 175 which is used for design. The slope of the creep-rupture curve of Groups I and II specimens is smoother than the scatter band curve. At low stress levels the rupture times fall within the scatter band (Fig. 9). This is generally observed for all creep-resistant

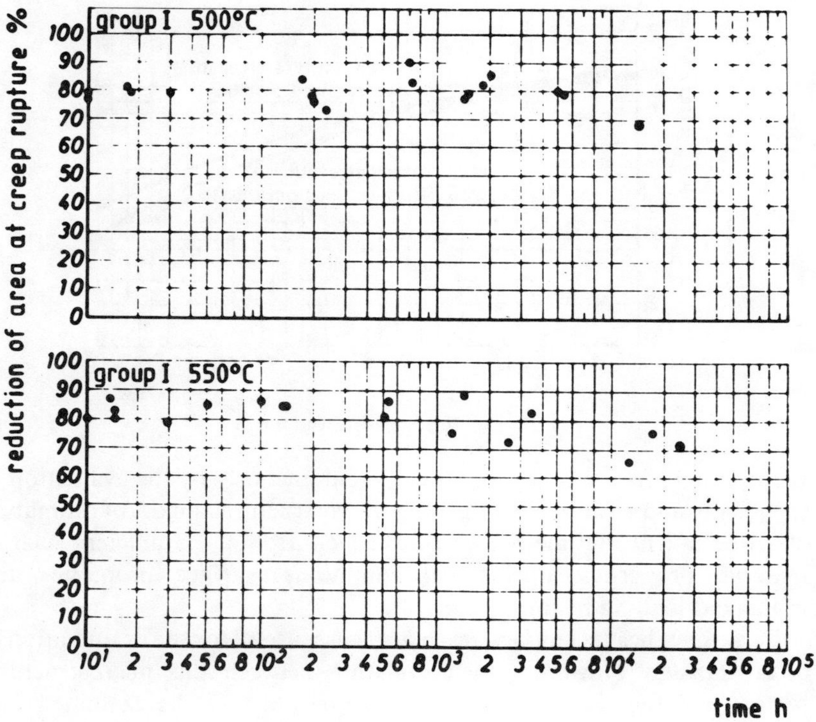

FIG. 10. Creep-rupture reduction of area of operational exposed components of 10CrMo9 10.

ferritic steels. The values of creep-rupture reduction of area (Fig. 10) are of the same magnitude as those of specimens tested in the as-received condition.

The following results can be concluded. The creep-rupture behaviour of the pre-exposed specimens cannot be estimated by the lifetime fraction rule derived from data of the above-mentioned scatter band. From the lifetime fraction rule a reduced creep-rupture strength has to be expected, even at low stress levels, which is not supported by experimental results.[20] The creep-rupture behaviour of a specimen cut from a 50% exhausted component can be derived as proposed in Fig. 11. The characteristic properties obtained from specimens reflect the influence of microstructural changes although the exhaustion is not assessed. The controlling mechanisms of structural change and exhaustion are not identical during operation; however, both phenomena occur. It is very difficult to predict

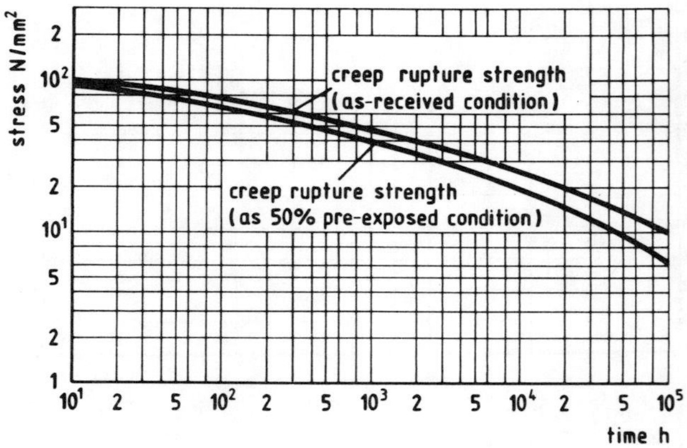

FIG. 11. Creep-rupture curves.

the residual lifetime of a component simultaneously by the evaluation of the locally and momentarily existing microstructure alone. An estimation of all remanent lifetime has to consider controlling parameters such as material properties, operational conditions, surface properties and multiaxial loading history.

Surface replica techniques are, in any case, useful for the examination of creep damage. However, the correlation between this microstructural feature and the residual mechanical properties must be examined very carefully. The estimations depend on:

(a) The kind of material.
(b) The exposure temperature.
(c) The kind of creep-controlling mechanism.
(d) The kind of crack-growth-controlling mechanism.

2.3. Transferability of Uniaxial Results to Components

The evaluation of stress intensity attempts to correlate component behaviour with the results obtained on uniaxially tested specimens. In general, each component is exposed to multiaxial loading conditions.

A tube, under internal pressure for instance, is stressed in three main directions: radial (r), circumferential (u), and axial (z). In cylindrical coordinates, the stresses are given by the stress tensor

$$\sigma_{ij} = \begin{pmatrix} \sigma_r & 0 & 0 \\ 0 & \sigma_u & 0 \\ 0 & 0 & \sigma_z \end{pmatrix}$$

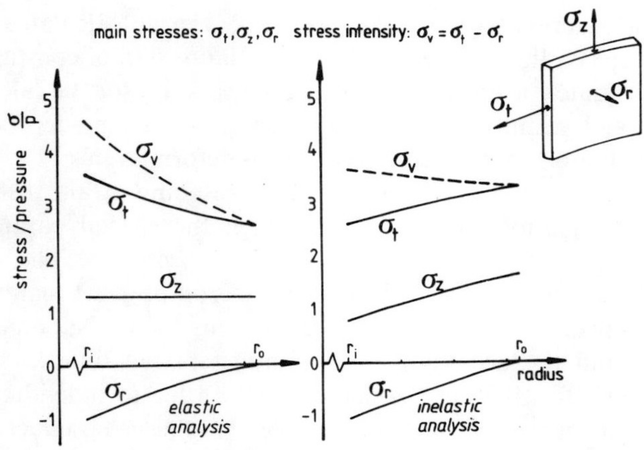

FIG. 12. Comparison of calculated stress distributions in a tube with 60 bar internal pressure by elastic and inelastic methods.

INCOLOY 800 H, T= 950°C, stress intensity σ_{eff} =30 N/mm^2		
specimen N°	loading condition	time to rupture
AYP 10	"dog bone", tension	112 hours
AYL 3B	tube, tension	217 hours
AYL 3C	tube, internal pressure	77 hours

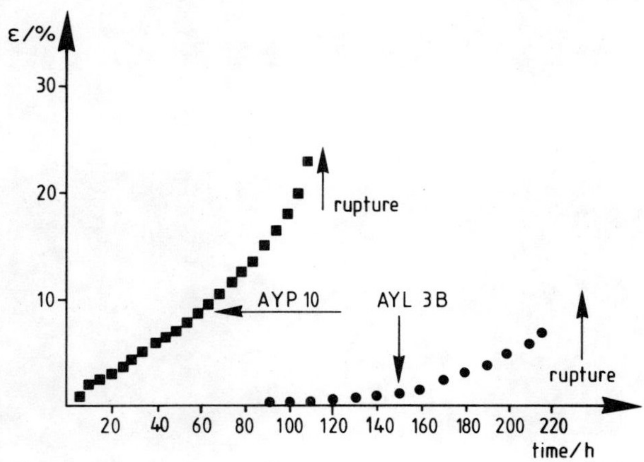

FIG. 13. Comparison of creep behaviour of a rod and a tube specimen of Incoloy 800 H. (σ_{eff} is the effective stress under multiaxial loading.)

The elastic reaction to this stress condition can be calculated (Fig. 12). Under creep conditions, the von Mises[21] relationship of constant volume and an adequate constitutive equation are necessary for the calculation of the stress and strain distributions, which are also time-dependent. The stress distribution is changed during creep deformation.

In general, for the evaluation of the stress and strain distributions, constitutive equations are used, for which the special parameters are derived from results on uniaxially loaded specimens. The specimens themselves, are frequently fabricated out of rod or sheet material. If the specimens are directly machined out of the component, their uniaxial creep behaviour and the necessary creep parameters can differ. This scatter results from differences in the microstructure due to individual thermo-mechanical treatment during processing the semi-finished products. To avoid this microstructural effect, differently shaped semi-finished products are produced from one master heat under the same specification with regard to grain size and final heat treatment. In Fig. 13 the creep behaviour at 950 °C of X 10 CrNiAlTi 32 20 (Incoloy 800 H) obtained on a specimen of rod

FIG. 14. Cracks in failed tubes: (a) Uniaxial creep test: 950 °C, $\sigma = 30\,N/mm^2$, time to rupture = 217 h. (b) Interpressure creep test: 950 °C, $\sigma_{eff} = 30\,N/mm^2$, time to leak = 77 h.

material and on an integral uniaxially tested tube are compared. The tube has the longer lifetime. A tube with such an internal pressure that the same stress intensity factor was reached failed earlier by leakage than would be expected from the rupture time of the tube under uniaxial loading. In the table of Fig. 13 the rupture time of a uniaxially loaded specimen is compared to the time to leak of the internally pressurised tube. The fracture appearances are shown in Fig. 14. Udoguchi and Nakanishi[22] have undertaken a series of creep experiments on tubes and on welded tubes of Hastelloy X under internal pressure. Typical creep curves derived from the increase of the diameter are given in Fig. 15. The results on the non-welded tubes (solid curves) demonstrate typical creep curves with secondary and tertiary stages. The welded specimen, however, expanded nearly linearly with the lower creep rate but failed before reaching the tertiary stage. The lifetimes of welded test pieces are significantly shorter and the range of rupture deformation much smaller than those of non-welded cylinders. In this experimental work, no attempt is made to prove theories or constitutive equations for the description of tube behaviour.

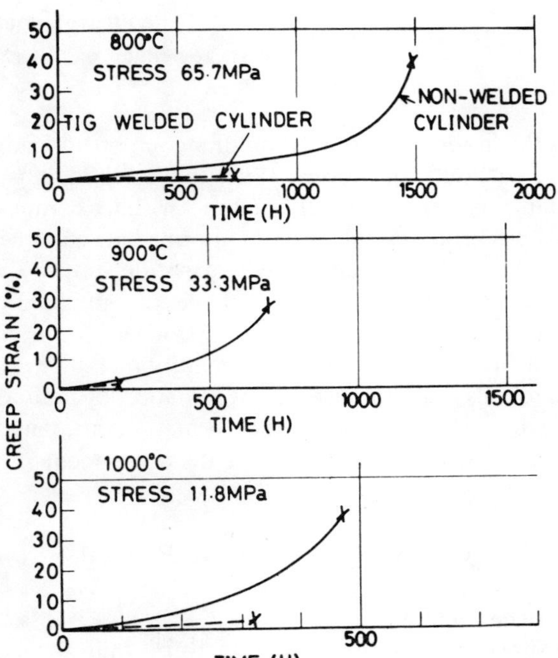

FIG. 15. Typical creep curves of the middle of a cylindrical specimen of Hastelloy X.[22]

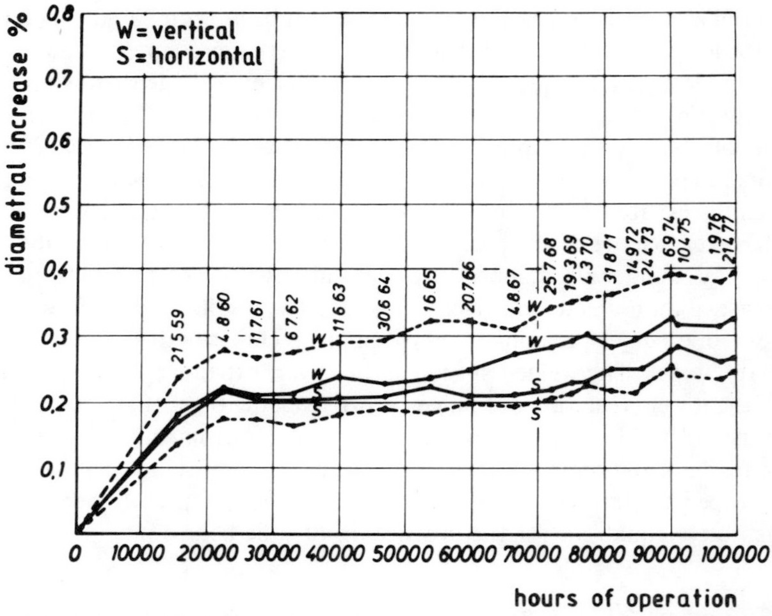

FIG. 16. Diametral deformation of superheater duct of 13CrMo44 during operation.

Measuring the in-service deformation of a component leads to the best understanding of multiaxial creep. With this method it is possible to directly evaluate the exhaustion of the material within a component. Figure 16 illustrates a practical example. For an intermediate superheater duct of 13 CrMo 44 the remaining diametral deformation during the operating time is plotted. If the component behaviour estimated by lifetime prediction analyses fulfills the obtained extension values during operation, the materials behaviour under multiaxial conditions can be described with the proper materials data and a precise estimation of multiaxial loading history.[20] The method to determine the deformation rate of a component is directly connected with the exhaustion of the components.

3. CONSIDERATION OF FATIGUE

3.1. Creep-fatigue Failure

In high temperature components, start-up, shut-down and power adjustments lead to thermal cycling which can introduce strain-controlled

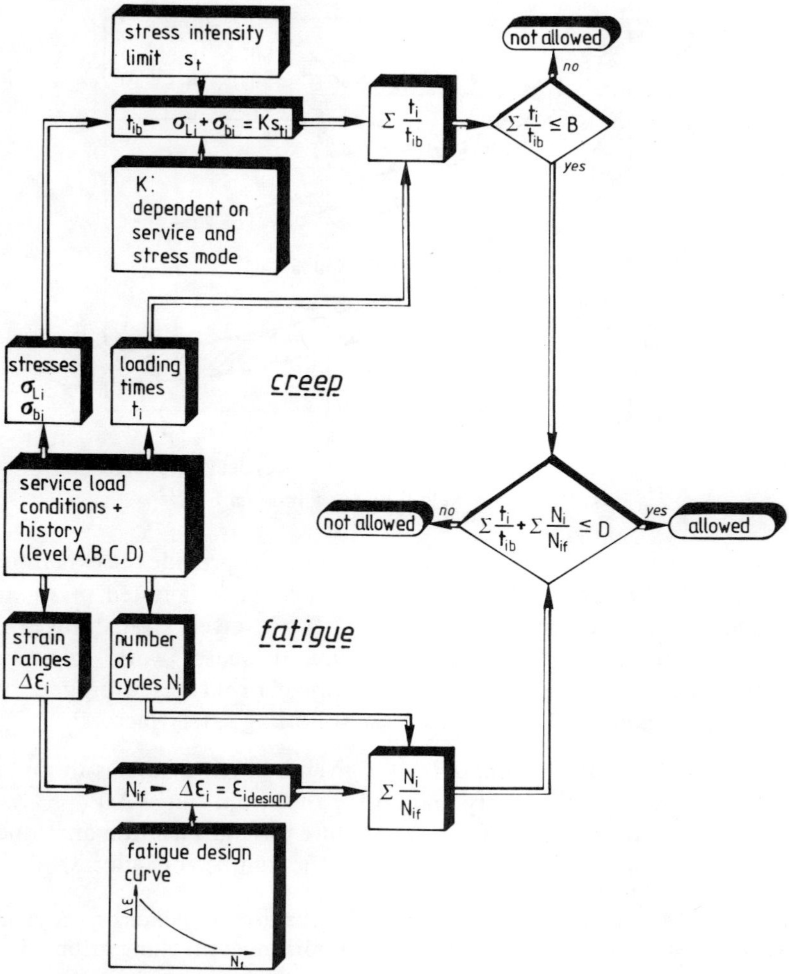

FIG. 17. Scheme of design against creep–fatigue interaction at elevated temperature according to ASME CC N47.

fatigue problems. For the assessment of these influences on component behaviour, low-cycle fatigue testing is required. For the analysis of a component, according to ASME Code Case N47 (Fig. 17), the consumption of allowable service lifetime is divided into pure creep and pure fatigue terms. For the stress analysis the acceptability of stress amplitude and the number of load cycles must be proved. A fatigue design curve in addition to the stress/time to rupture curve is needed.

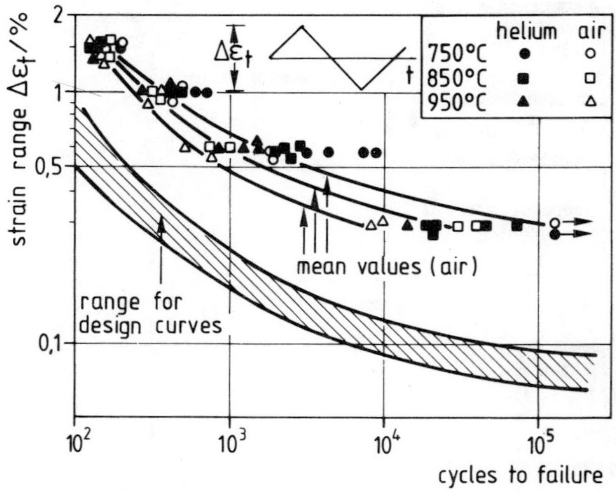

FIG. 18. Low-cycle fatigue of Inconel 617.[22]

Low-cycle fatigue tests at different temperatures provide the experimental basis. In Fig. 18 results on the alloy Inconel 617, tested in air and simulated HTR helium, are collected.[23] For a given strain range, the endurable number of cycles to failure decreases with increasing temperature, especially in the lower strain range. From these average curves a design curve is obtained by using the following safety factors:

$u_{\Delta\varepsilon} = 2$, for the strain amplitude at a given number of cycles to failure (where $u_{\Delta\varepsilon}$ is the safety margin for the strain range $\Delta\varepsilon$)

$u_N = 20$, for the number of cycles to failure for a given strain amplitude (where u_N is the safety margin for the number of failures, N)

In the high temperature region the fatigue behaviour depends on the strain rate; this is why creep becomes the most important controlling mechanism. With decreasing strain rate, the number of endurable cycles decreases (Fig. 19).[24] This is especially true for low values, where the strain rate is more characteristic of component behaviour. In turbine component design the beginning of crack initiation is preferred to the number of cycles to failure criterion. The procedure of the experimental determination of both values implies some uncertainties. Both values depend on crack initiation and crack propagation. Crack propagation seems to be creep-controlled at very high temperatures. Microstructural fracture analysis after test temperatures above 800 °C indicates that the crack morphology seems to be caused by creep rather than by fatigue.

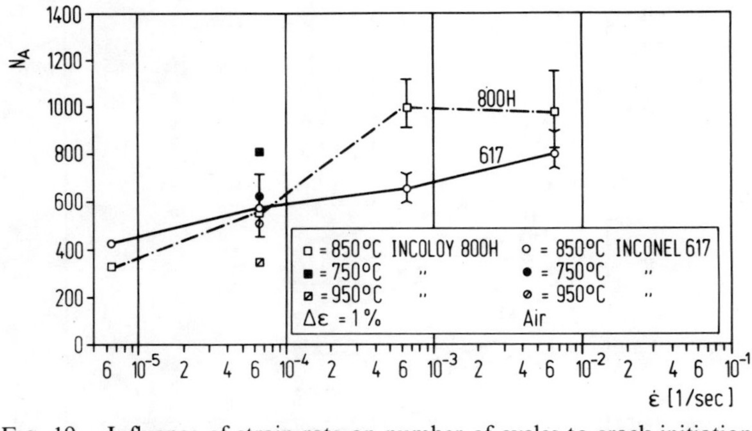

FIG. 19. Influence of strain rate on number of cycles to crack initiation.

With increasing hold time, the number of cycles to failure is reduced (Fig. 20).[23] Fatigue design curves for the evaluation of the results of elastic analysis should take into account the influence of strain rate and of hold time. The number of cycles to failure on specimens tested in simulated helium is higher than obtained on air-tested specimens. This is a clear indication of the impact of corrosion on fatigue life.

For the evaluation of the results of inelastic analysis, an accurate assessment of the stress condition is recommended. Fatigue curves can thus be used for the evaluation of results of inelastic analysis without hold time.

3.2. Creep-fatigue Accumulation

Besides the linear accumulation rule proposed by ASME CC N47 some other analytical methods for the correlation of creep and fatigue are known.[25] Manson *et al.*[26] treat the accumulated creep strain and the time-independent plastic strains in tension and compression as damage (strain range partitioning method—SRPM). As described before, this work favours the definition of the damage D as the initiation and growth of microstructural voids and cracks. In this definition, the behaviour under cyclic strain-controlled loading is mainly controlled by creep mechanisms. It is very important, however, that this treatment is restricted to high temperatures and to limited strain rates.

The linear damage accumulation rule is the most common for the designer. However, linear accumulation using results of LCF tests with hold times must be very carefully examined with respect to the allowable total damage value D. From hold-time experiments[23] on Incoloy 800 H at

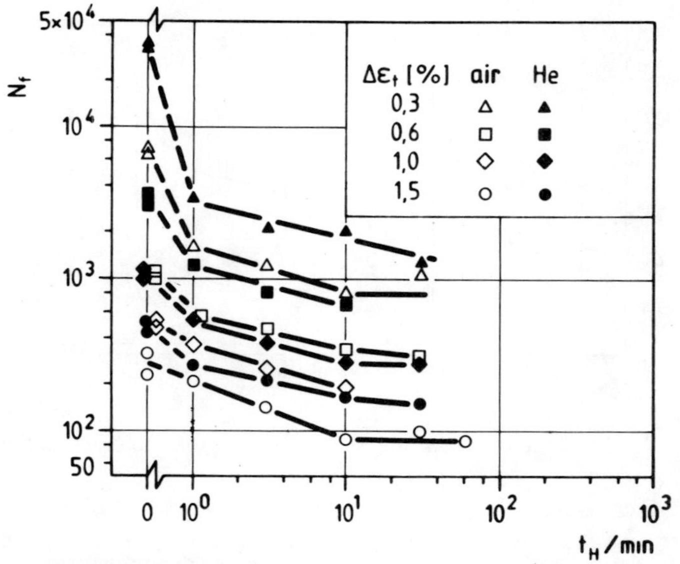

FIG. 20. Influence of hold time and test atmosphere on number of cycles to failure of Incoloy 800 H at 850 °C.[22]

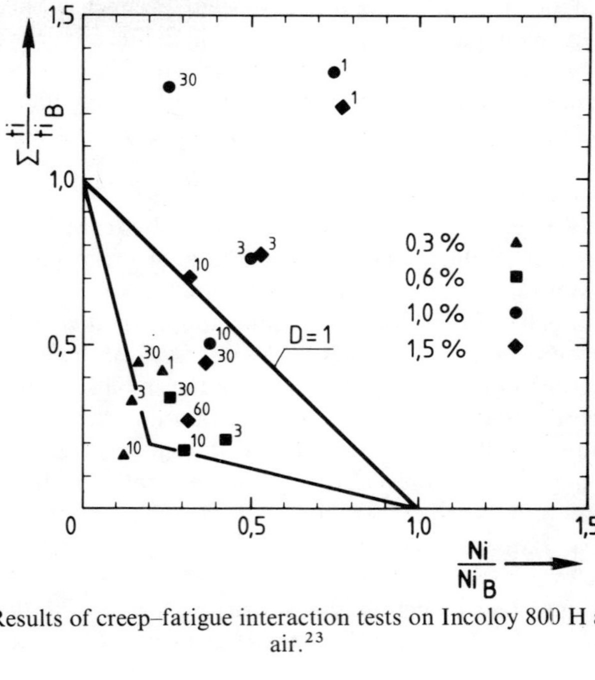

FIG. 21. Results of creep–fatigue interaction tests on Incoloy 800 H at 850 °C in air.[23]

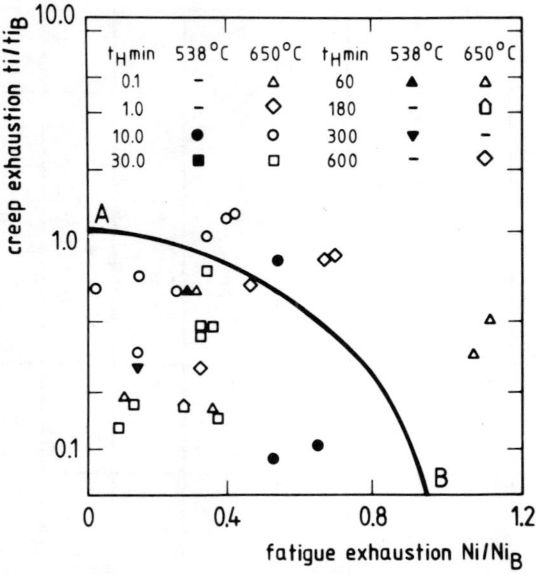

FIG. 22. Creep–fatigue interaction graph for AISI 304.[27]

850 °C fraction of time in creep versus fraction of numbers of cycles is plotted (Fig. 21). Some test results are below $D = 1$. In Fig. 22 a creep–fatigue interaction diagram according to Jaske *et al.*[27] is given for AISI 304. Some results are above $D = 1$, but the bulk of the data is far below. This may be a reason for ASME CC N47 to propose a bilinear approximation. MPA Stuttgart[28] performed extensive experimental work (Fig. 23) to prove the proposal of Timo.[29] Here the hold times are simply treated as creep exposure time. In fatigue tests with small strain ranges the time to cracking falls in the scatter band of time to rupture of the creep tests. From tests under different cyclic loads with hold times and from tests with packages of creep and cyclic loads, adapting coefficients may be obtained for each individual steel to describe the accumulation of lifetime sufficiently precisely. A generalisation, however, cannot be made. This is caused by the different development of the microstructure under the same stress or strain-rate conditions of each alloy. The thermomechanical history significantly affects microstructure of all high temperature components in power plant and gas turbines. Again it must be checked whether the material behaviour of specimens from the exposed component represents exhaustion of the component and whether severe change in number of cycles to failure is observed. In Fig. 24 some results are given where the obtained number of

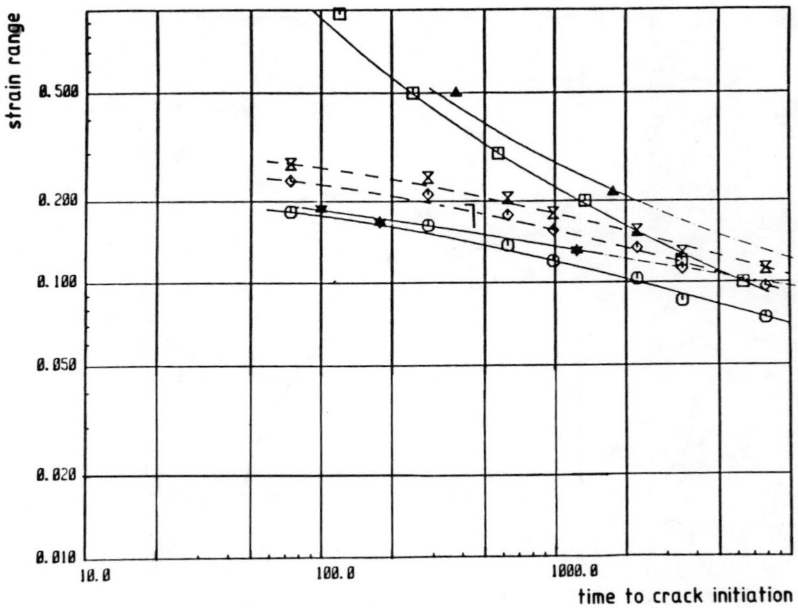

FIG. 23. Comparison of creep and fatigue test results on 13CrMo44 steel according to Timo.[29] (t_{hz} is holding time in tension, t_{hd} is holding time in compression.)

	Temperature (°C)	$\dot{\varepsilon}$ (% min)	t_{hz} (min)	t_{hd} (min)
◻	530	6·0	20·0	20·0
○	530	0	0	0
▲	530	6·0	20·0	20·0
◇	530	0	0	0
⧖	530	0	0	0
★	530	0	0	0

cycles to failure seems to be independent of exhaustion of service life. The linear accumulation rule does not describe the material behaviour within the component under multiaxial loading conditions.

For increasing temperature, the endurance limit for high cycle fatigue disappeared (Fig. 25) as demonstrated by results on Nimonic 86.[23] For the turbine blade material, under the framework of the COST 50 programme,[30] a series of high cycle fatigue tests has been done. The effect of

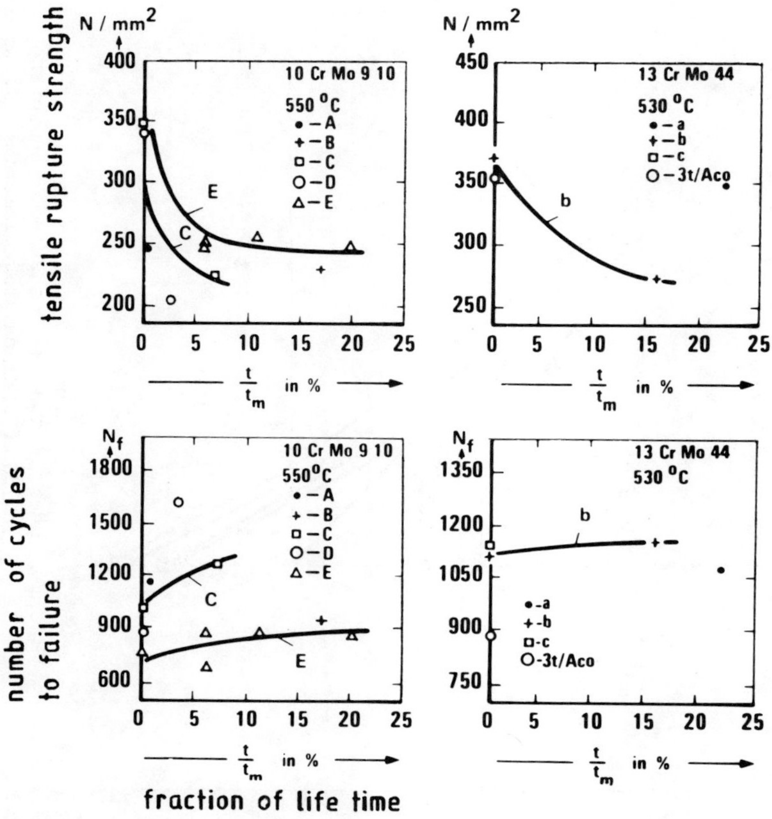

FIG. 24. Fatigue properties of specimen of service-exposed components.

mean stress on fatigue strength with increasing temperature is given in Fig. 26 and normalised to the 200-h creep-rupture data of smooth specimens. The creep strength reduction with increasing temperature is higher than the value of high cycle fatigue endurance at about 10^8 cycles.

High cycle loadings, and vibrations in the component induced by external events, below yield strength or 0·1 % proof strength can be handled with elastic analysis methods. The temperature limit is dependent on the material.

The influence of high cycle loading on lifetime of a component is a specific problem for turbine blades. There is some progress in tests with packages of high cyclic loads and creep exposure,[31] but the results have not yet been evaluated and published.

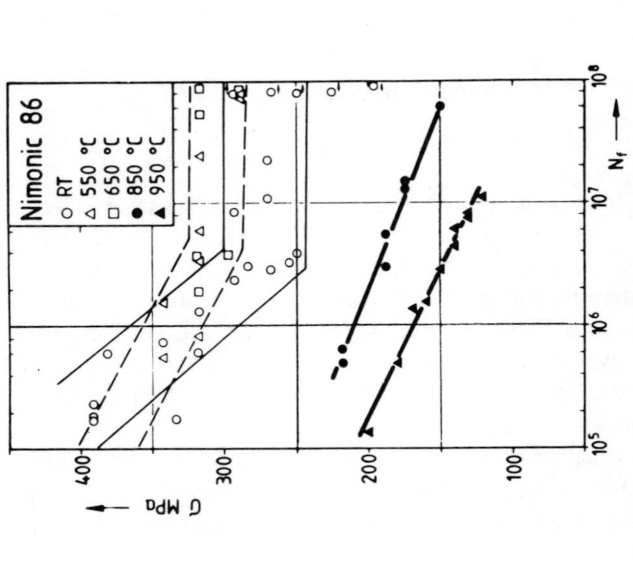

FIG. 25. Influence of test temperature on HCF results in Nimonic 86 (N_f is number of cycles).

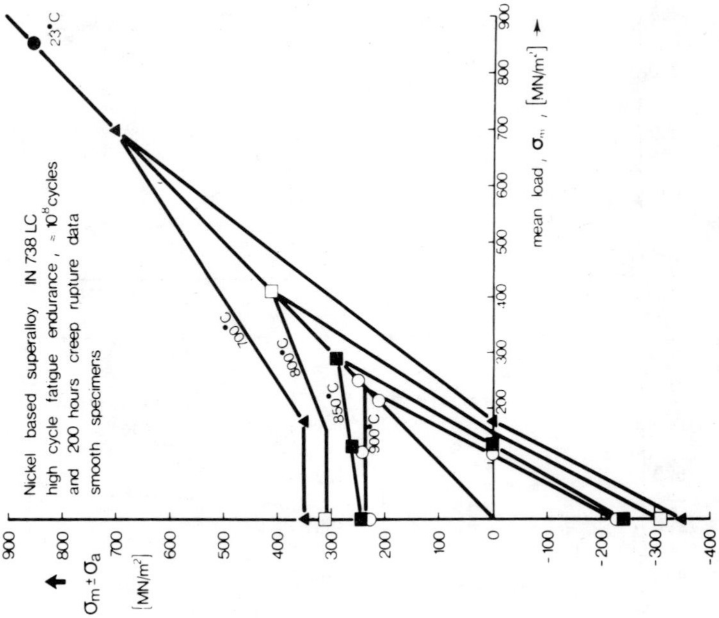

FIG. 26. Effects of mean stress on fatigue strength of alloy IN 738 LC at various temperatures.[30]

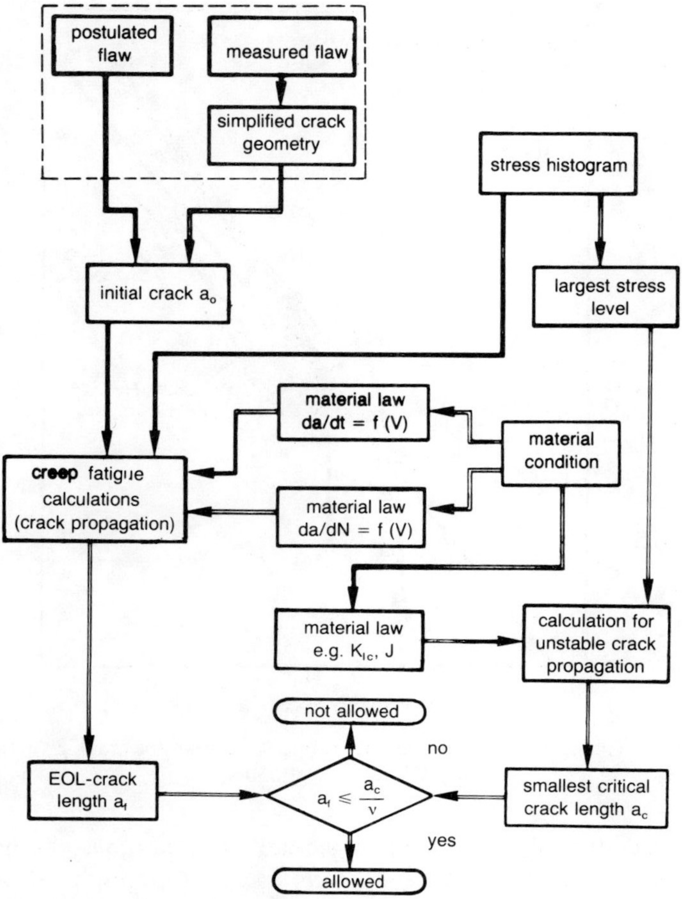

FIG. 27. Proposal for fracture mechanics analysis in the high temperature region.

3.3. Fracture Mechanics Evaluations

For components under pressure at room temperature and slightly elevated temperatures, fracture mechanics arguments are used for the evaluation of the behaviour of components with flaws. At higher working temperatures, where material properties are time-dependent, no established rule exists for the use of fracture mechanics methods. For components in the THTR circuit, a kind of Battelle formalism[32] has been applied as an approximation to ensure leak before break.

Figure 27 shows a scheme of a fracture mechanics analysis necessary for

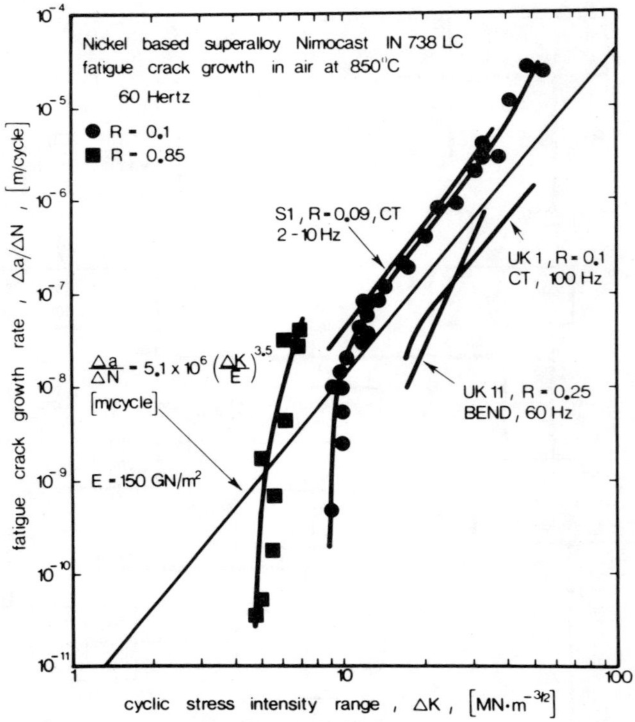

FIG. 28. Effect of cyclic stress intensity range ΔK on fatigue crack growth rate in alloy IN 738 LC at 850 °C.[30]

high temperature. By using suitable fracture mechanics concepts for cyclic crack propagation, da/dN, and creep crack propagation, da/dt, the propagation of a detected or postulated flaw is calculated and its size at the end of life determined. The method of fracture mechanics analysis shown in Fig. 27 is derived from the analysis proposed in ASME Code Section III, Appendix G, but for high temperature applications there is no valid fracture mechanics concept to describe crack propagation.

For HTR components of advanced systems the work on fracture mechanics has begun, but to date the establishment of a design rule has not been possible. A working group of BBC Baden[30,33] published results on crack propagation of gas turbine blade material. The dependence of fatigue crack growth rate (Fig. 28), da/dN, on the stress intensity range, ΔK, at 850 °C is similar to room-temperature behaviour. After a threshold value the growth rate increases rapidly before reaching a steady-state growth rate

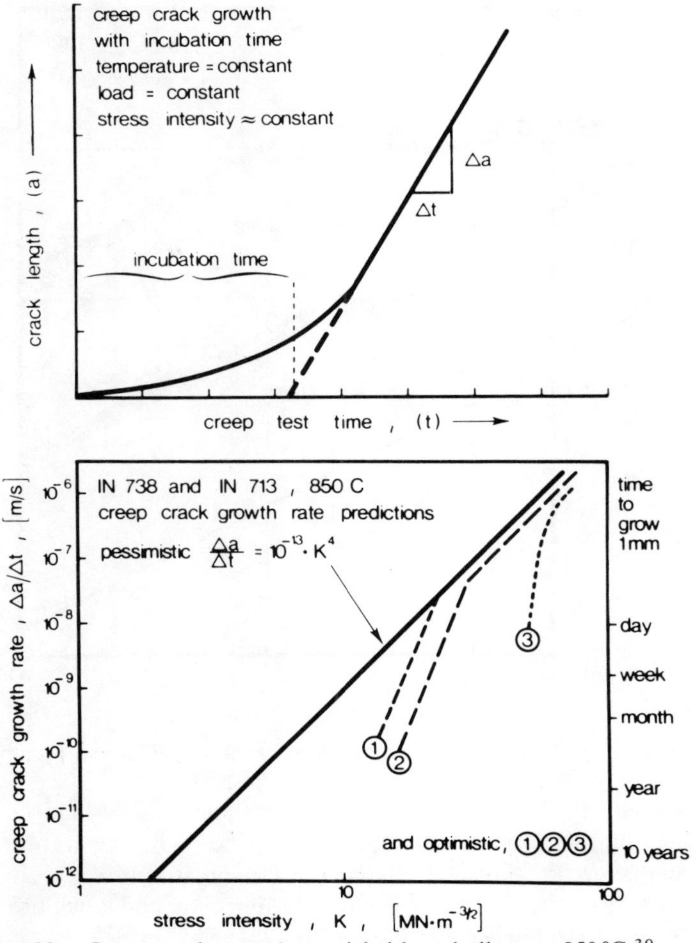

FIG. 29. Creep crack growth on nickel-based alloys at 850 °C.[30]

which then turns into another rapid increase up to failure. For the steady growth a kind of Paris correlation can be derived.

In some creep crack growth tests an incubation time for the crack growth is observed (Fig. 29) and has to be considered. At higher values of both parameters, creep crack growth rate, da/dt, and stress intensity factor, ΔK, an approximation like the Paris equation is possible, and thus a calculation of flaw length can be made. However, not all creep fracture experiments are sufficiently developed that a rule or an argument for the designer can be given. Speidel[30] (Fig. 30) proposes, in a simplified approach, to link

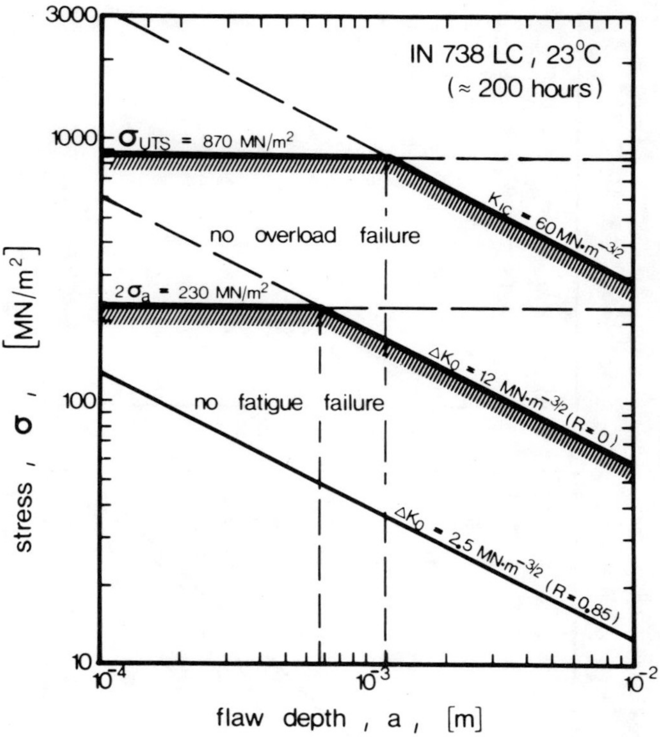

FIG. 30. The fracture mechanics approaches (Speidel and Pinau[30]).

stresses and flaw sizes to define areas safe against overload failure and fatigue failures. These approaches to describe high temperature fracture mechanics do not reflect the change of microstructure due to loading history and/or gas/metal reaction. Crack initiation and crack propagation can be greatly influenced by corrosion.

To deal with high temperature fracture mechanics the following features are needed, but are not yet available:

(a) A law for cyclic crack propagation for different microstructural conditions.
(b) A law for creep crack propagation for different microstructural conditions.
(c) An evaluation of high temperature limit values.

Crack initiation and growth depend on the multiaxial stress situation ahead of a crack tip, but at higher temperatures the stress may relax.

4. CONSIDERATION OF STRUCTURAL DEFORMATION

The primary stresses can be superimposed by a stress due to temperature gradient across the wall of a component. The secondary stress, which can be preliminarily estimated by elastic analysis, relaxes with the time at temperature. In Fig. 31[34] the tangential stress distribution through a tube wall with a temperature gradient across the wall, after relaxation, is calculated by the Norton creep law for increasing service durations. Under the condition of stationary operation the stress relaxation causes deformation of the wall. The problem becomes more difficult to treat when shut-down procedures are superimposed.

A particular class of failure results from progressive deformation. Localised deformation can occur by the superimposition of primary and secondary stresses. The deformation can accumulate until a crack is initiated which may finally lead to component failure if it is repeated during further cycles, e.g. at each start-up procedure. By limiting the stress to $3 \times S_m$, according to ASME Code Part NB or the KTA rules, the incremental collapse is avoided in the range of temperature where material properties are time-independent. ASME Code Case N47 also requires a ratchetting analysis for the range of time-dependent properties. Four criteria are recommended. The first two test criteria ensure that the material loading in a single load fluctuation remains elastic. This is fulfilled when the

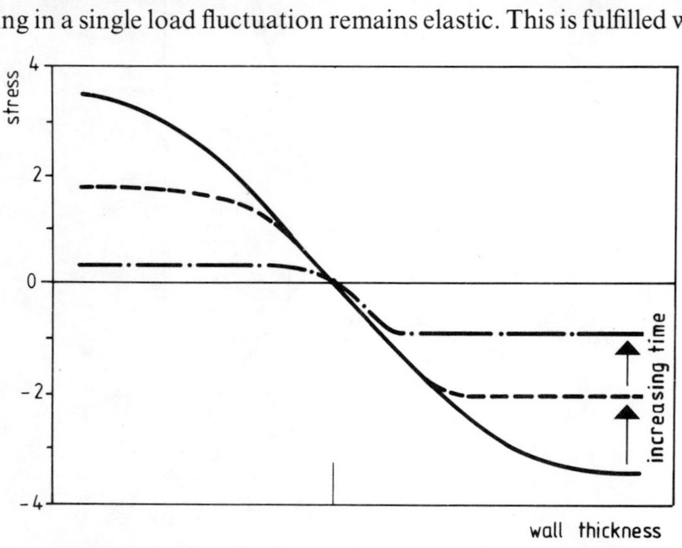

FIG. 31. Calculated stress distribution in a tube wall with a temperature gradient across the wall.[34]

secondary stress at the lower temperature wall side is low and the sum of maximum primary and secondary stresses does not exceed the 0·2 % proof stress. In the third test the total strain limits are calculated by an equivalent creep stress; this is the sum of the maximum primary and secondary stress intensities each multiplied by 1·25. This sum must be within given limits. Criterion four seems to be fulfilled when all accumulated strains are below the 1 % creep limit.[35]

4.1. Creep Ratchetting

Ratchetting is defined as progressive cyclic inelastic deformation in a component. It is a superimposition of mechanical secondary stresses, thermal secondary stresses or both on the sustained primary stress. For a cylindrical component Bree[36] and Anderson[37] have made some calculations under the conditions of constant internal pressure and variable temperature gradient across the wall. Using elasto-plastic materials behaviour they defined regions for allowable primary and secondary stresses (Figs 32 and 33):

(a) *Region E*: the cross-section is cyclic loaded within the pure elastic range.

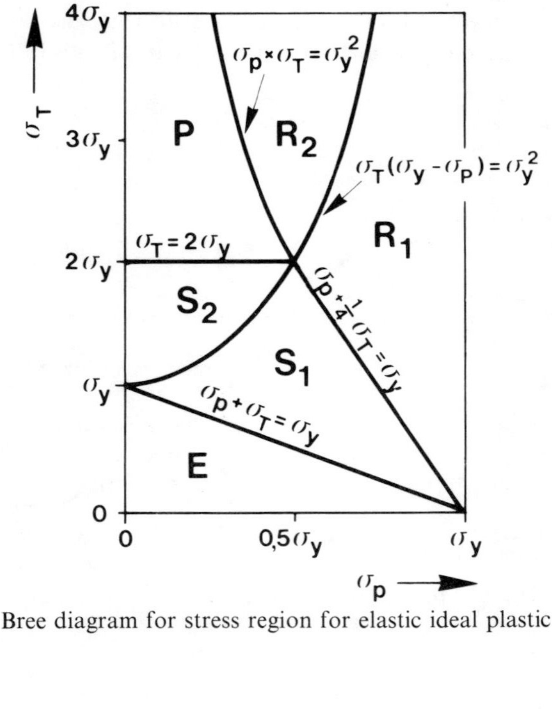

FIG. 32. Bree diagram for stress region for elastic ideal plastic materials.

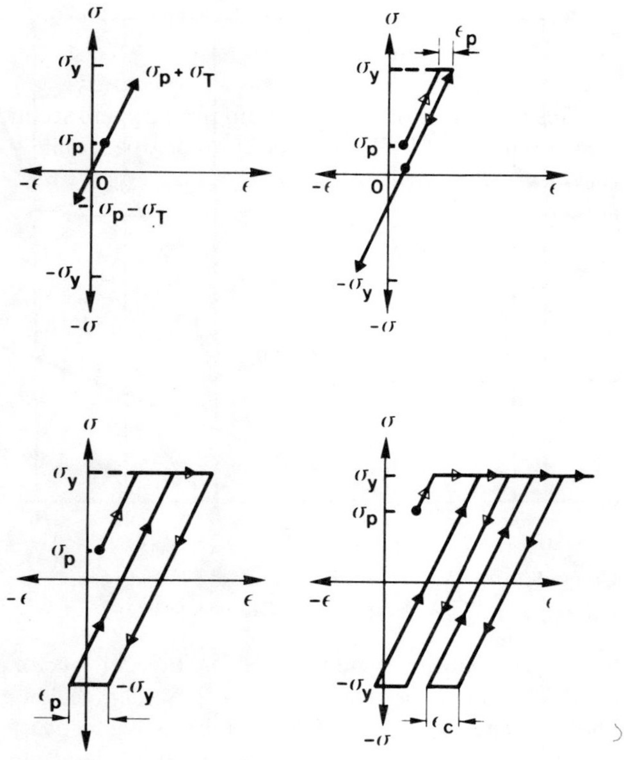

FIG. 33. Stress–strain curves in dependence of primary to secondary stress.

(b) *Regions* S_1 *and* S_2: the plastic flow during the first cycle has to cause hardening to such an extent that the following fatigue loading remains in the elastic range.

(c) *Regions* R_1 *and* R_2: plastic flow at one side of the tube wall induces stresses at the other side. Thus with constant thermal cyclic loading the material deformation increases within each cycle by the same order of deformation.

(d) *Region P*: these areas represent a strain fatigue exposure.

For exposure within the creep region the values are normalised, but with a simplified model[38] the behaviour under creep strain conditions can be estimated. Within regions R_1 and R_2, the total creep strain (Fig. 34) is successively compared to pure creep under primary stress level. A series of

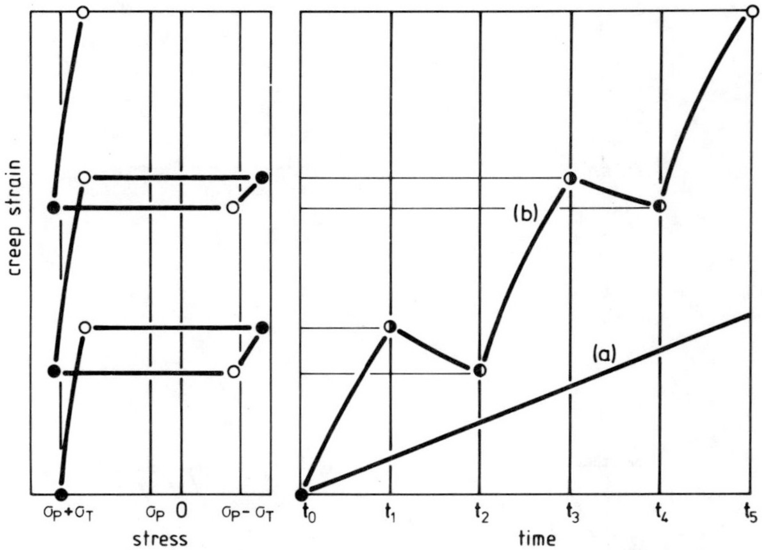

FIG. 34. Creep deformation in dependence of stress and time.[37] (σ_p is primary stress, σ_T is temperature induced stress.)

experiments at the operating conditions of fast breeder reactors has been performed, which allows the definition of a boundary below which the creep ratchetting effects can be estimated by an 'effective stress' line (Fig. 35).[37] This effective stress leads, on average, to the same remanent creep strain as is expected for the superimposition.

An application of the ASME Code Case N47 test is restricted to medium temperatures. For higher temperatures the development of new simplified methods is necessary.

The example discussed for creep ratchetting shows that for components operating under stress at high temperatures an increase in permanent strain is unavoidable. For strain analysis, those positive main inelastic strains which are a result of the loading history are ensured against strain limits (Fig. 36). According to ASME Code Case N47, the limits for positive inelastic strain are:

(a) 1 % for strain averaged through the thickness.
(b) 2 % for surface strains due to an equivalent linear distribution of strain throughout the wall thickness.
(c) 5 % for local strain at any point.

The strain limits for weldments are set at 50 % of the above values.

The strain criteria cannot be quantitatively related to a specified failure

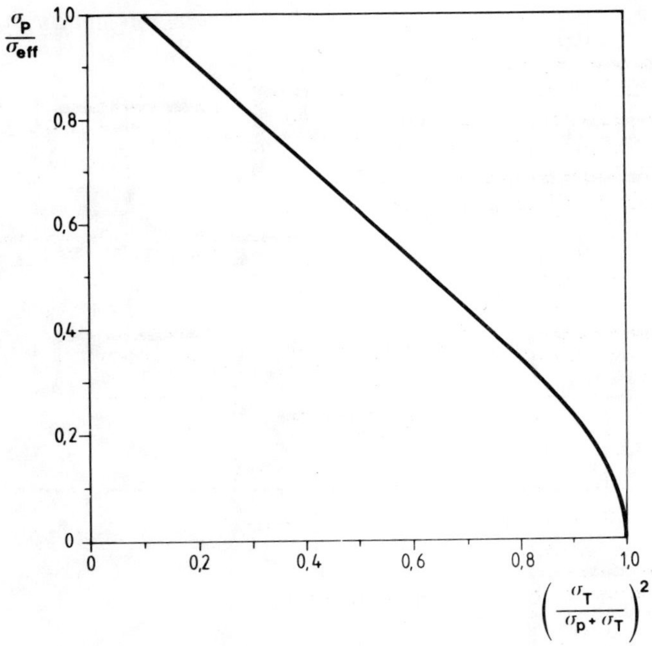

FIG. 35. Experimental developed maximum curve for an effective stress.[37]

mode but they allow the definition of further quantified estimates with reference to the following failure modes:

(a) Creep rupture due to long-term loading.
(b) Creep-fatigue failure.
(c) Excessive strain due to incremental deformation or ratchetting.

The strain criteria are a means of controlling all deformation occurring during the operation of a component. The total strain is a function of the loading history and reflects the interaction of different loading conditions and microstructural changes. The strain criterion does not differentiate between strain due to plasticity and strain resulting from creep.

Effects due to gas/metal reactions at the surface are not incorporated into these strain limits.

4.2. Loss of Function

The increasing strains can effect a loss of function. Examples are:

(a) The loosening of flange connections due to stress relaxation in the bolts.

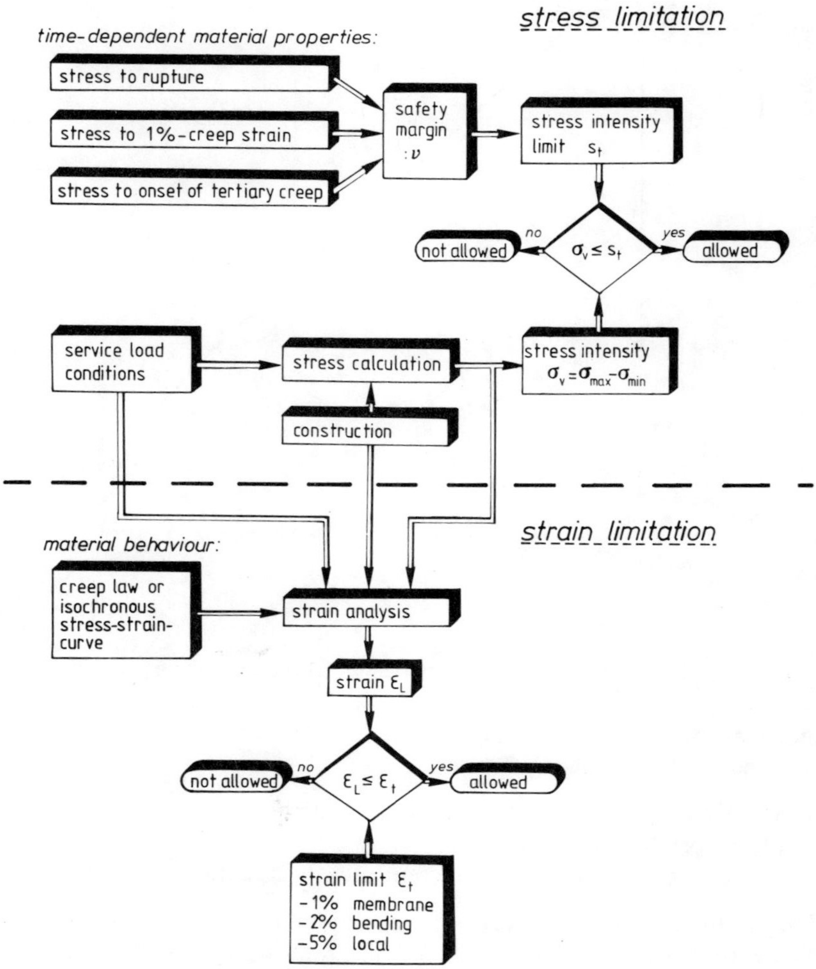

FIG. 36. Scheme of stress and strain limitations at elevated temperature according to ASME CC N47.

(b) The loosening of valves due to creep deformation of components under pressure.

(c) The change in tolerance or gaps due to different creep deformation behaviour of parts of a component.

The requirements for function may lead to more restrictive stress or strain limits, as mentioned previously.

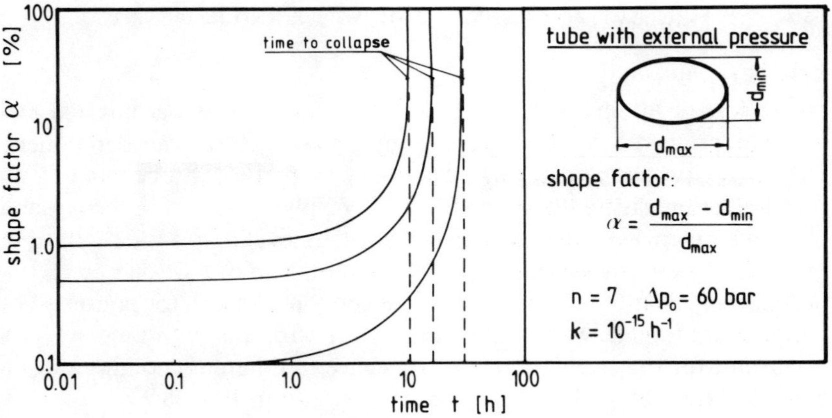

FIG. 37. Creep buckling in a tube.

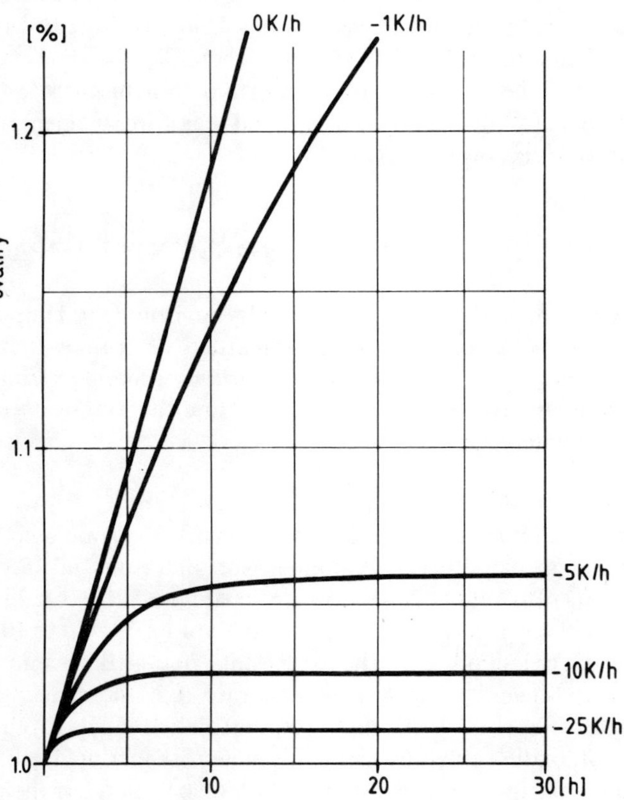

FIG. 38. Influence of cooling rate on creep buckling.[39]

4.3. Creep Buckling

The loss of internal pressure in a tube component under internal and external pressures causes stresses, which may deform the component. Dependent on temperature and materials properties, a tube exhibits an increase of ovality with time (Fig. 37). Applying Hoff's[39] theory and Norton's creep law, the expected change in ovality for three starting ovalities is given. In the calculation, the exponent n of Norton's creep law has an important influence on the creep buckling deformation. The temperature level as well as the temperature variation with time are very important for the calculation. The influence of different cooling rates on creep buckling of an HTR component is shown in Fig. 38.[35,40]

ASME Code Case N47 differentiates between time-independent and time-dependent loss of stability. Whereas the time-independent loss of stability is characterised by kinking, tilting or buckling, the time-dependent loss of stability is a problem of peak loading. Peak loading causes increased deformation due to creep or relaxation. Therefore the time under these peak loads must be restricted. It is important to demonstrate that time-dependent loss of stability cannot occur, and therefore sufficient time safety margins have to be supplied.

5. FAILURE DUE TO ENVIRONMENTAL IMPACT

The above failure modes can be enhanced by environmental impact. There are working atmospheres where corrosion attack determines lifetime. As a result, mechanical property data are increasingly generated from samples tested in simulated working atmospheres. Here the long-term properties are of special interest.

5.1. Creep–Corrosion Interaction

For many applications, creep experiments in simulated working atmospheres are in progress. A comparison of results of specimens of different heats of Incoloy 800 H,[41] in air, in simulated HTR helium and in simulated methane reforming gas are plotted in Fig. 39. Up to 10 000 h, all results fall in the same scatterband. Results obtained in simulated gas turbine atmospheres[42] (Fig. 40) demonstrate that the remaining cross-section without corrosion effects is responsible for creep rupture life. A correction of loading stress by using this 'net cross-section' leads to values which fall within the scatterband of air-tested specimens. If the gas/metal reaction leads to internal microstructural changes, e.g. carburisation or selective corrosion, creep behaviour and rupture life may be affected.[43] The drastic reduction in rupture life observed in Ref. 42 can be explained by the reduction of load-carrying cross-section due to material wastage and internal sulphidation. A change in the fracture mode from intergranular to transgranular has been reported.[42]

Constant strain-rate testing methods (Fig. 41) have been used to determine a critical strain rate above which detrimental creep–hot corrosion interaction will occur. A specimen of IN 738 LC exposed in a synthetic slag has been examined by Grünling *et al.*,[42] but no critical strain rate could be established. This was caused by the severe corrosion with high sulphidation effects. A significant reduction in flow stress and rupture elongation was observed for a specimen under slag compared to specimens tested in air. The dotted line represents the calculated curve when the loss in

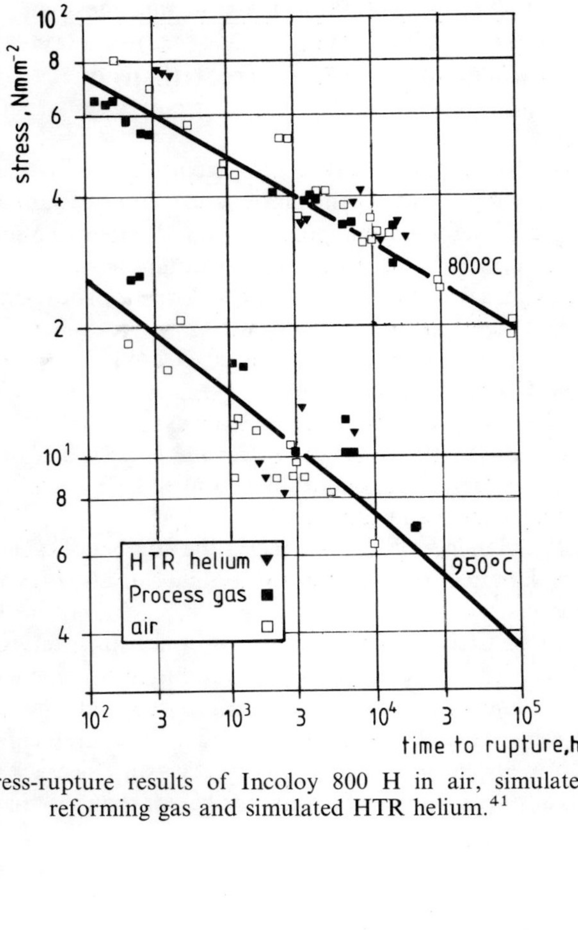

FIG. 39. Stress-rupture results of Incoloy 800 H in air, simulated methane reforming gas and simulated HTR helium.[41]

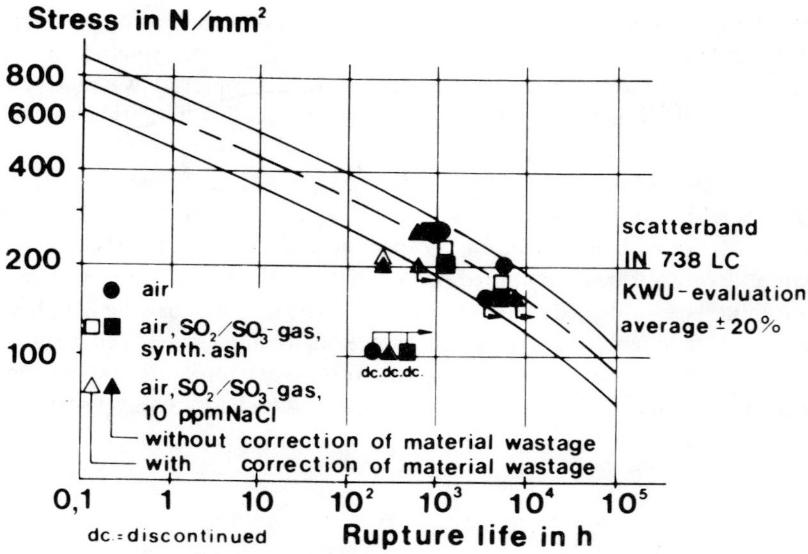

FIG. 40. Stress-rupture test results of IN 738 LC at 850 °C.

load-bearing cross-section is taken into account. This reduction was not formed to the same extent for the higher corrosion-resistant alloy IN 939. The influence of sulphidation processes on the deformation behaviour could not be verified by microstructural examination.

For the design, a wall-thickness margin still seems to be the only practical approach to dimensioning a component.

5.2. Fatigue–Corrosion Interaction

The number of cycles to failure depends not only on mechanical properties but also on the status of the surface. Crack nucleation processes may have an influence on endurable number of cycles, and therefore corrosion influences the fatigue lifetime.

In vacuum and in inert gas atmospheres the number of cycles to failure is often higher than in corrosive atmospheres such as air (Figs 18 and 20) or hot burning gas.

There is a restricted amount of work concerning fatigue–corrosion interaction at high temperatures. Some fatigue tests[44] have been carried out on specimens in synthetic ash; both as-received and pre-corroded specimens were used. Fatigue crack growth measurements have been performed[33] in air, in vacuum and in synthetic ash. The crack propagation rate in air is about one order of magnitude higher than that in vacuum. No

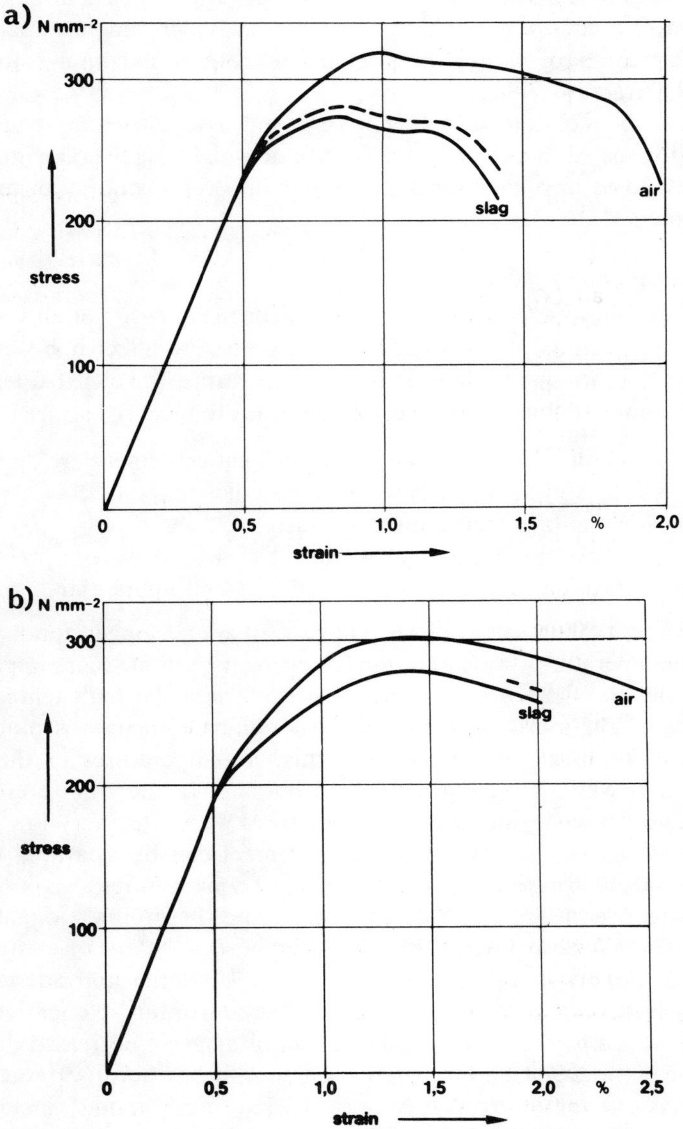

FIG. 41. Stress–strain diagrams for: (a) IN 738 LC, and (b) IN 939 from constant strain-rate tests in air and corrosive slag.[20,21] $T = 850\,°C$; $\dot{\varepsilon} = 10^{-8}\ s^{-1}$.

difference was observed between crack propagation rates in air and under sulphidising conditions. Chlorine in the test environment supported a transformation of the crack mechanism from transgranular to intergranular fracture.

Alterations of surface morphology and microstructure caused by corrosion may enhance creep crack initiation and fatigue crack initiation. This has to be accounted for in design for wall cross-section transmittance and notches.

5.3. Irradiation Effects

The metallic components used in a nuclear reactor are not only affected by the operational environment but may also be affected by fast and thermal neutron irradiation. At room temperature and elevated temperatures the interaction of fast neutrons with the atoms of the lattice leads to:

(a) Irradiation-induced hardening and embrittlement by agglomerations and interstitials which form voids and clusters.
(b) Swelling of material due to voids.
(c) Irradiation-induced creep due to stress-induced preferential absorption (SIPA effect) of vacancies and interstitials.

Above about 500 °C these effects recover and are no longer dominant. At higher temperatures transmutations of elements become more important. Especially the life from (n, α)-reaction influences the high temperature ductility of the metals utilised. Irradiation-induced helium is practically insoluble in metals and migrates at higher temperatures to the grain boundaries where it agglomerates into bubbles. In the grain boundaries these bubbles cause intercrystalline fractures with often very low plastic deformation. The extent of the embrittlement can be measured by hot tensile and creep testing. The influence of test temperatures—an important parameter for this kind of embrittlement—and the rupture elongation of an austenitic steel are given in Fig. 42. It can be seen that the microstructure plays an important role in the ductility.[45] While the unirradiated and thermally aged samples feature an increased rupture elongation with increasing temperature, the irradiated samples have a decreased ductility between 600 and 900 °C. The better irradiation behaviour of variant KA 2, which has a fine grain size, may be due to carbide precipitating preferentially within the grains and being an effective sink for helium. That the embrittlement may be successfully influenced by special thermomechanical treatments of the alloy is an encouraging indication.

Irradiation-induced property changes have to be considered for design

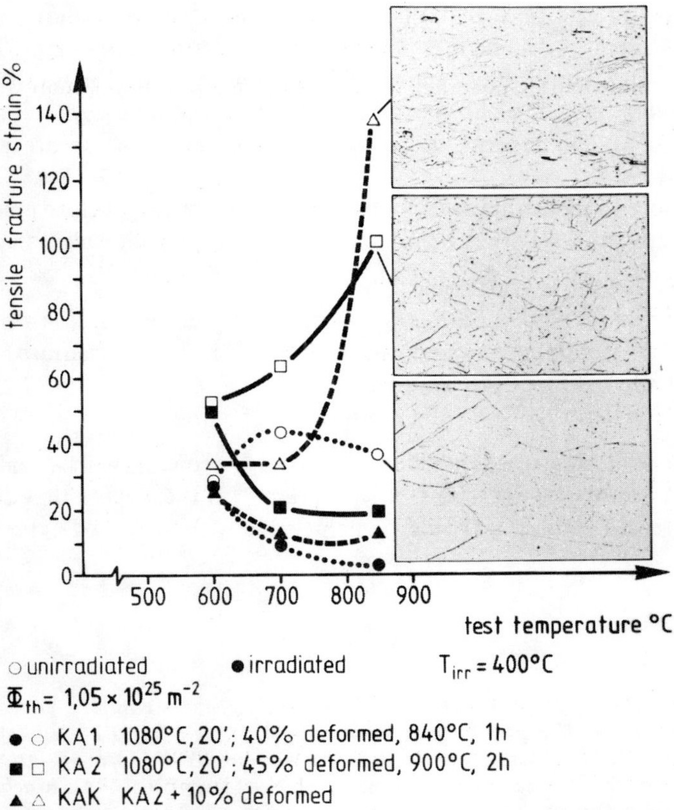

FIG. 42. Influence of microstructure of post-irradiation fracture ductility of stainless steel 1·4981. (Φ_{th} is neutron influence.)

by an adequate safety margin, especially where ductile behaviour is concerned.

6. CONCLUSIONS

The property interactions and multiaxial problems discussed are illustrated by experimental results obtained mainly on specimens.

Tests on high temperature components are not too common in materials evaluation programmes, except for experiments in rigs for gas turbine components.[46] Examination of entire pressure vessels and pipings for PWR nuclear power stations by testing under simulated loading and evaluation of the deformation behaviour and kind of failure ensures reactor safety.[47,48] Some component tests have also been performed on zircalloy cladding of fuel elements.[49] Component tests for non-nuclear fossil-heated power plant and the petrochemical industry are as yet unusual. For components of advanced HTRs experiments are still in progress.

The more economical use of the potential of high temperature materials leads to research activities being undertaken on the following topics:

(a) Constitutive equations.
(b) Lifetime prediction methods and damage accumulation rules.
(c) Transferability of materials data laws which better adjust for the multiaxial loading condition.
(d) Non-destructive testing methods.

The issue of lifetime prediction and multiaxial approaches will be dealt with as a high priority by the COST 501 program.[50] It is expected that designs with increased economy should be achievable with increased experience.

REFERENCES

1. Metallische Werkstoffe für HTR (1982). Statusseminar Düsseldorf, Schriftenreihe *Energiepolitik in Nordrhein-Westfalen*, Band 14, Minister fuř Wirtschaft, Mittelstand und Verkehr des Landes NRW (14 & 15 Jan.).
2. KFA-Bericht über den Forschungszeitraum Nov. 1979–Dec. 1982, *Erarbeitung von Grundlagen zu einem Regelwerk über die Auslegung von HTR-komponenten für Anwendungstemperaturen oberhalb 800°C*.
3. Coutsouradis, D., Felix, P., Fischmeister, H., Habraken, L., Lindblom, Y. and Speidel, M. O. (eds) (1978). *High Temperature Alloys for Gas Turbines*, (COST 501 Conference 1978, Lüttich), Applied Science Publishers, London.
4. Brunetaud, R., Coutsouradis, D., Gibbons, T. R., Lindblom, Y., Meadowcroft, D. B. and Stickler, R. (eds) (1982). *High Temperature Alloys for Gas Turbines*, (COST 50 Conference 1982, Lüttich), D. Riedel Publishing Company, London.
5. Weber, H. (1983). *Anl. RWTÜV-Seminar, Auslegung, Herstellung und Betrieb von zeitstandbeanspruchten komponenten* (April).
6. AD-Regelwerk, Werkstoffe von Druckbehältern, Reihe W; Berechnung von Druckbehältern, Reihe B; Schwellbeanspruchung, Reihe S.
7. TRD-Regelwerk, Allgemeine Grundsätze für Werkstoffe, Reihe 100; Festigkeitsberechnung von Dampfkesseln, Reihe 300.
8. DIN 2413, Stahlrohre, Berechnung der Wanddicke gegen Innendruck.
9. KTA-Regel, Komponenten des Primärkreises von Leichtwasserreaktoren, Teil: Auslegung, Konstruktion und Berechnung, 3201.2.

10. ASME Code, Case Section III, Part NB, Class 1, Components.
11. ASME Code (1979). Case N47, Section III, Division 1, Class 1, Components in elevated temperature service.
12. Norton, F. H. (1938). *Creep of Steels at High Temperatures*, McGraw Hill, New York.
13. Kachanov, L. M. (1967). *The Theory of Creep*, Nat. Lending. Lib. for Science and Technology, Yorkshire, England.
14. Robinson, E. L. (1952). Effects of temperature variation on the long time rupture strength of steels, *Trans. ASME*, **74**, 777–80.
15. Mordfin, L., Halsey, L. and Green, G. E. (1959). NASA TN 0-181.
16. Barth, H., Granacher, J., Kloos, K. H. and Rieth, P. (1983). Hochtemperaturverhalten warmfester Stähle unter zeitlich veränderter Beanspruchung. *Anl. VGB-Conference, Werkstoffe und Schweißtechnik im Kraftwerk 1983*, Düsseldorf (Jan.).
17. Jakobeit, W., Pfeifer, J. P. and Ullrich, G. (1982). Ertüchtigung von Turbinenwerkstoffen. Statusseminar Düsseldorf, Schriftenreihe *Energiepolitik in Nordrhein-Westfalen*, Band 14, Minister für Wirtschaft, Mittelstand und Verkehr des Landes NRW (14 & 15 Jan.).
18. Penkalla, H. J., Over, H. H. and Schubert, F., Constitutive equations for the description of creep and creep rupture behaviour of metallic materials at temperatures above 800 °C, *Nuclear Technol.*, to be published.
19. Brenner, U., Neubauer, B. and Schulze, H. D. (1983). Bewertung von Fehlern in kriechbeanspruchten Kraftwerksbauteilen. *Anl. VGB-Conference, Werkstoffe und Schweißtechnik im Kraftwerk 1983*, Düsseldorf (Jan.).
20. Weber, H. (1981). Service life analysis of high temperature installation under creep conditions. In: *Int. Symp. Prediction of residual life time of constructions operating at high temperature*, The Hague.
21. Von Mises, R. (1913). *Mechanik der festen und flüssigen Körper im plastisch deformablen Zustand*, Königliche Gesellschaft der Wissenschaften, Göttingen.
22. Udoguchi, T. and Nakanishi, T. Structural behaviour of a welded superalloy cylinder with internal pressure in a high temperature environment. *Int. J. Pres. Ves. & Piping*, to be submitted.
23. Meurer, H. P., Gnirß, G., Mergler, W., Raule, G. and Schuster, H. (1982). Untersuchung zum Ermüdungsverhalten von HTR-Werkstoffen bei Temperaturen bis zu 1000 °C. Statusseminar Düsseldorf, Schriftenreihe *Energiepolitik in Nordrhein-Westfalen*, Band 14, Minister für Wirtschaft, Mittelstand und Verkehr des Landes NRW (14 & 15 Jan.).
24. Schuster, H. and Schiffers, H., KFA-results. *Nuclear Technol.*, to be published.
25. Miner, A. A. (1945). *J. Appl. Mech.*, **12**, 159.
26. Manson, S. S., Halford, G. R. and Hirschberg, H. H. (1971). Creep-fatigue analysis by strain range partitioning, NASA TMX-67838.
27. Jaske, C. E., Mindlin, H. and Perrie, J. S. (1972). ASTM, STP 520, 365.
28. Bhongbhibhat, S. (1979). Untersuchungen über das Werkstoffverhalten im Gebiet der Zeitstandfestigkeit zur Erstellung von Berechnungsunterlagen für überwiegend thermisch beanspruchte Bauteile. *Techn. wiss. Bericht der MPA-Stuttgart*, Heft 79-02.
29. Timo, D. P. (1969). Designing turbine components for LCF. *Int. Conf. on Thermal Stress and Thermal Fatigue*, Berkeley (Sept.).

30. Speidel, M. O. and Pinau, A. (1978). *High Temperature Alloys for Gas Turbines* (COST 501 Conference 1978, Lüttich) (eds D. Coutsouradis, P. Felix, H. Fischmeister, L. Habraken, Y. Lindblom, and M. O. Speidel), Applied Science Publishers, London.
31. Jakobeit, H. private communication from BBC-Mannheim.
32. Bodmann, E. private communication from HRB-Mannheim.
33. Hoffelner, W. and Speidel, M. O. (1982). The influence of the environment on fatigue crack growth of the nickel-based superalloys IN 738 LC and IN 979 at 850 °C. *ASME-Power Conf.*, Denver, 82-JPGC-NE 1 (Oct.).
34. Breitbach, E. (1982). KFA-Jül.-Spez. 154, Mechanische und thermische Betriebsbelastungen der Rohre eines nuklear beheizten Röhrenspaltofen (June).
35. Over, H. H. (1984). KFA internal report, published in the final report of *Erarbeitung von Grundlagen zu einem Regelwerk über die Auslegung von HTR-Komponenten für Anwendungstemperaturen oberhalb 800 °C*, Jül.-Spez. (Manh), p. 248.
36. Bree, J. (1968). Incremental growth due to creep and plastic yielding of thin tubes subjected to internal pressure and cyclic thermal stresses. *J. Strain Anal.*, **3**(2).
37. Anderson, W. F. (1971). Creep ratcheting deformation and rupture damage induced by a thermal transient stress cycle. *First Nat. Congr. on Pres. Ves. and Piping Technol.*, San Francisco (May).
38. Schubert, F., Over, H. H. and Zottmaier, R. (1983). Darstellung des Kriechratchetings. *Anl. RWTÜV-Seminar, Auslegung, Herstellung und Betrieb von zeitstandbeanspruchten Komponenten* (April).
39. Hoff, N. J., Jahsman, W. F. and Nachbar, W. (1959). A study of creep collapse of a long circular cylindrical shell under uniform external pressure. *J. Aero-Space Sci.*, **26**(10).
40. Schubert, F., Seehafer, G. and Bodmann, E. (1982). Status of design work in Germany concerning materials and structural aspects for the heat exchanger components of advanced HTR's. *ASME-Power Conf.*, Denver, 82-JPGC-NE 1 (Oct.).
41. Schubert, F., Bruch, U., Cook, R., Diehl, H., Jakobeit, W., Penkalla, H. J., teHeesen, H. and Weber, H. (1982). Zeitstandverhalten. Statusseminar Düsseldorf, Schriftreihe *Energiepolitik in Nordrhein-Westfalen*, Band 14, Minister für Wirtschaft, Mittelstand und Verkehr des Landes NRW (14 & 15 Jan.).
42. Grünling, H. W., Keienburg, K. H. and Schweitzer, K. K. (1982). The interaction of high temperature corrosion and mechanical properties of alloys. *High Temperature Alloys for Gas Turbines* (COST 50 Conf. 1982, Lüttich) (eds R. Brunetaud, D. Coutsouradis, T. R. Gibbons, Y. Lindblom, D. B. Meadowcroft and R. Stickler) D. Riedel, Publishing Company, London.
43. Bürgel, R. (1981). Dissertation Zeitstandverhalten der Stähle X 10 NiCrAlTi 32 20 und X 50 NiCrAlTi 33 20 in Luft und aufkohlender Atmosphäre.
44. Grünling, H. W. and Schneider, K., Private communication.
45. Thiele, B. A., Weber, H., Ohly, W. and Diehl, H. *Nuclear Technol.*, to be published.

46. Hansen, W. and Esslinger, P. (1978). *Werkstofftechnische Probleme bei Gasturbinentriebwerken*, Werkstofftechnische Verlagsgesellschaft mbH, Karlsruhe.
47. KTA-Fachseminar (1979). Integrität und Festigkeit druckführender Reaktorkomponenten.
48. MPA-Seminar (1979, 1980, 1981, 1982). *Sicherheit de druckführenden Umschließung von Leichtwasserreaktoren*, Stuttgart (Oct.).
49. Leistikow, S. and Schanz, G. (1979). H. v. Berg: KfK-Bericht 2810.
50. COST 501—European Concerted Action (1982). *High Temperature Materials for Conventional Systems of Energy Generation and Conversion Using Fossil Fuels.*

Index